Enterprise Release Management

Agile Delivery of a Strategic Change Portfolio

For a complete list of titles in the
Artech House Technology Management and Professional Development Series
turn to the back of this book.

Enterprise Release Management

Agile Delivery of a Strategic Change Portfolio

Louis J. Taborda

ARTECH
HOUSE

BOSTON | LONDON
artechhouse.com

Library of Congress Cataloging-in-Publication Data
A catalog record for this book is available from the U.S. Library of Congress.

British Library Cataloguing in Publication Data
A catalogue record for this book is available from the British Library.

Cover design by Irene Ehrmann

ISBN 13: 978-1-60807-168-5

10 9 8 7 6 5 4 3 2 1

Contents

Preface

Anyone who has worked in a modern enterprise knows it is a complex environment where attempts to mobilize resources to achieve strategic outcomes can be frustratingly difficult. With differing and often conflicting agendas of business units, project teams, and IT departments, it can sometimes appear as if the greatest challenge in large organizations come not from competitors but from within the enterprise itself.

Business executives, who have the unenviable task of charting an enterprise's strategic course through an increasingly competitive, globalized marketplace, recognize that agility and adaptability can provide competitive advantage. Yet the reality is that executing a strategy and getting ideas to market expediently and cost-effectively can be problematic. The clamor internal to the enterprise complicates decision making and the market response is slowed by internal inefficiencies including poor communications and information sharing. Priorities get confused between the plan and its execution, so that what is delivered to customers is often a compromise that is late and over-budget.

As the engine room of the modern enterprise, information technology (IT) is at the heart of many initiatives that require the aggressive delivery of the capabilities needed to execute the business strategy. However, delivering these capabilities—the products and services demanded by the business within time and cost constraints—has often proved too great a challenge. The failure of IT projects has been a common theme in recent times and disappointment with IT's ability to deliver on its potential has become a prevailing sentiment in the enterprise, irrespective of whether it is an in-house department or outsourced service provider.

Yet these are mere symptoms of the general lack of clarity evident in the modern enterprise, and projects are simply the bellwether, or the messenger, highlighting our growing inability to cope with enterprise complexity. A recent survey of global CEOs conducted by IBM [1] found that while today's complexity is only expected to rise, more than half for the over 1,500 CEOs, general managers, and senior public sector leaders around the world interviewed doubted their ability to manage. These leaders emphasized the need for creativity, experimentation, and innovation, suggesting that we are at a juncture where earlier management truths are breaking down without the opportunity to fully comprehend why. It is no surprise therefore, that many business leaders have an uneasy feeling that once trusted approaches and practices are somehow less relevant.

This book aims to identify the root causes of the problems executing strategic programs and presents a remedy we term enterprise release management (ERM)—a release-centric management framework for reliably delivering strategic outcomes. Not all readers of this book may experience that Eureka! moment where the problem crystallizes and the challenges of planning and delivering strategic change in complex environments are resolved. But I am confident that the ideas herein will spark new insights and offer practical and effective ways to synthesize the multiplicity of competing demands that have to be juggled in successfully delivering an enterprise change portfolio.

Scope and Terminology

An indication of the newness of a paradigm is the difficulty in expressing it. Finding the appropriate language to describe ERM and the generalizations it entails present some challenges. Many words that one must rely on to describe the release paradigm, including the word release itself, are vague, overloaded with several meanings, or commonly corrupted. It is particularly difficult to establish generalizations without the temptation to coin a completely different language.

An attempt has therefore been made throughout the book to use generic terms that have minimal baggage, and sometimes that has meant going back to a more simple language. Introducing terms like solution delivery into the discourse have required consideration, as to simply use program/project management (PPM) terminology would not be appropriate when arguing that a new release paradigm was in fact needed.

The word delivery is appropriate, as many projects today do not require development, being either implementations of packaged solutions or involving outsourced teams that are not under direct project control. As such, it is hoped that the increasing awareness of services delivery and management terminology and its straightforward interpretation provide the flexibility for ERM to encompass more than just development, and indeed transcend the project construct as a delivery vehicle. Without relying on general terms, we would be in the paradoxical situation of using old management paradigms to describe new ones; a situation that can obscure the fundamental issues that have to be addressed.

Similarly, the use of the word solution is an attempt at generality. While much of the language describing the enterprise comes from the IT industry, use of the word solution instead of software, system, or application attempts to recognize two things:

1. That there is more to any business change than simply delivering new technologies; processes and people aspects have to be considered, along with the financial means necessary to achieve the desired outcome.
2. Solutions can also be seen to encompass products and services of indeterminate type, including marketing, facilities, and engineering initiatives that may form a part of an enterprise strategy.

Solution is therefore used to broaden the scope of what a release may constitute, and like the term project, can be seen to be generic and independent of what it is that is being delivered. While the word system in its most general meaning might equally suffice, common usage tends to reduce a system to integrated hardware and software rather than interpret it in a wider organizational context.

Even the term management brings with it preconceptions and is open to interpretation. The road to ERM is one that takes us from the command and control interpretation of traditional management to that of greater collaboration and coordination required in managing decentralized teams and structures. The tools and techniques that arise from the release paradigm can be used to attain a new level of governance and oversight, or they can be applied to improve coordination between self-managed teams. ERM itself is neutral and agnostic of the more detailed methods and process frameworks; encompassing both plan-driven and agile approaches to delivery. It offers a unifying governance framework without imposing its own management philosophy or demanding that traditional management approaches be abandoned. Instead it provides a notation for synthesizing and aggregating crucial management information at a suitable level of abstraction that allows leaders to make informed and timely decisions.

Who Should Read This Book

The tussle with terminology described above and the attempt to keep the language neutral, avoiding the biases of any particular discipline is worth the effort to ensure readers from a cross section of the enterprise can find common ground. Keeping this book accessible and understandable to the different enterprise roles is essential as there is little chance of integrating the different facets of the federated enterprise if we cannot find a common representation and framing of our collective experiences.

Presenting a new, unifying paradigm for enterprise evolution needs to bring together the perspectives of those responsible for the determination, planning, execution, and realization of the optimum business strategy mix. This book therefore targets different stakeholders across the "enterprise release life cycle," offering different insights to each while maintaining a coherent vision that aims to bring these factions (dare one say silos) together. Without getting lost in the detail of each specialization, this work targets practitioners and students in the following key readership groups:

- *Business and IT executives* who sponsor the strategic initiatives and want to shorten the delivery timescales and have greater management visibility of progress. Such roles are looking to deliver business value and avoid the frustrating delays that impede the implementation of their strategies.
- *Portfolio managers and project management office (PMO) managers* who work across multiple programs and projects to provide consolidated reporting. These central roles need to reconcile and prioritize often conflicting demands in order to best allocate limited resources.
- *Program/project manager, product managers and business analysts* who collect and analyze customer requirements to flesh out the strategy and drive the solution design, but suffer compromised delivery schedules when facing the reality of limited resources.
- *Enterprise and solution architects* who are responsible for aligning business and IT strategy and identifying solution alternatives, while ensuring business and IT impacts are understood across the increasingly complex and interdependent enterprise architecture.
- *IT and technology managers* responsible for the delivery of the technical components—whether systems, products, or services that have to be procured, developed, engineered, or integrated—to provide the solution capabilities. As creators of the enabling technologies that ultimately power the enterprise, they feel the pressure of working to often unrealistic budgets and schedules.
- *Testers and delivery managers* who, as the last step between the plans and the reality, have to consolidate multiple delivery streams into a coherent release, identify stakeholders and customer proxies who are impacted, and ensure the final release is accepted and transitioned into business as usual.

- *Business and IT operations* who are responsible for maintaining the business-as-usual (BAU) but have their own changes and enhancements that can get sidelined as the focus is on major projects that consume resources and management attention.

Observant readers will note the omission of the release manager role. This role has been purposefully avoided due to the many interpretations the role has in the enterprise—something that this book hopes to address and refine. So to those of you already acting in that capacity, this book can offer the means to extend your influence and clarify interactions with related roles in the enterprise.

Structure of the Book

Attempting to present a grand unification of enterprise change and transformation is one thing; navigating the results is another. The problem with introducing a new, somewhat heretical, and possibly disruptive management paradigm is that to describe it needs the reader to step back and reexamine some fundamental assumptions.

Yet in our time-poor world we will often just scan a book, quickly find the key concepts, and move on. So in an attempt to make the content digestible, this book has a modular structure which, like any good architecture, should allow the reader to focus on a particular aspect of the ERM story. The structure of this book, shown in Figure P.1, provides a guide to the main concepts which are organized as follows:

- The first chapter makes the case for release management at the enterprise level. ERM should be firmly grounded in the shared management concerns related to project execution, systems development, and enterprise architecture.
- The next two chapters look at the contrasting approaches to solution delivery which currently hold sway. Chapter 2 reviews the two apparently opposed management viewpoints of plan-driven versus agile development that respectively address change and complexity. Chapter 3 then discusses the ubiquitous project paradigm, identifying conditions in the enterprise that can reduce the effectiveness of project management. While these chapters revisit arguments that have been made elsewhere, they provide the point of departure from more familiar management approaches to increasingly necessary release-centric approaches.
- The crux of the ERM is described in the next two chapters. Chapter 4 identifies the enterprise pattern that inevitably results when multiple

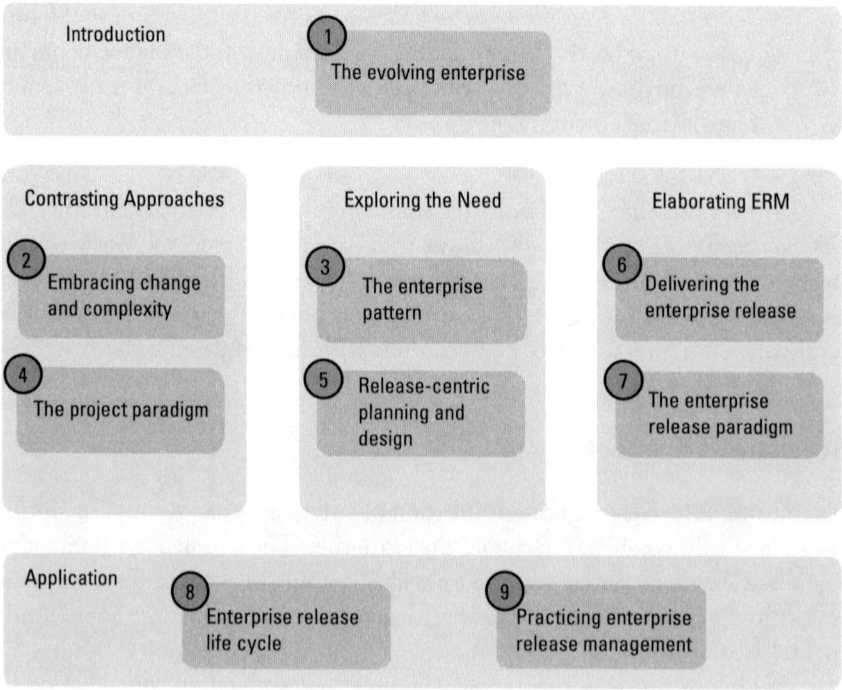

Figure P.1 Navigation for readers.

projects impact a complex enterprise architecture. The resulting interdependencies between these initiatives give rise to systems behavior that are shown to need more integrated management. Chapter 5 identifies the release as an important factor in selecting a change portfolio for execution. These chapters highlight the need for integrative and cohesive techniques to balance the potential multiple changes have to splinter an architecture.

- ERM practices are elaborated in the subsequent chapters with Chapter 6 relating the release-centric perspective to an industry study on business alignment versus IT efficacy. This motivates an analysis of the different relationships between business customers and solution suppliers patterns that give rise to corresponding delivery patterns. Chapter 7 introduces the matrix notation that resolves the most complex enterprise pattern and offers a representation for the multiple stakeholder viewpoints present in an enterprise release.
- The final two chapters apply ERM principles with Chapter 8 extending the matrix model to provide a multiproject release life cycle model. A timeboxed release roadmap is shown to offer a framework for planning

and executing strategy. Chapter 9 discusses the implementation of ERM in the enterprise and discusses the implications of the new approach and its relationship to different practices in the enterprise.

To get the most from this book, it is recommended that the reader at least read the introduction of each chapter sequentially even as the remaining details can be selectively skipped for future consideration. Figure P.1 outlines the particular areas where the different roles can expeditiously concentrate.

Reference

[1] Institute of Business Value, *Capitalizing on Complexity*, IBM Corporation, 2010.

Introduction

This book proposes the release as the next paradigm for the management of evolving businesses and IT architectures. Past efforts to improve efficiency by breaking down silos and consolidating investments in IT solutions now mean that changes to the business are no longer isolated. Instead, growing interdependencies mean that even simple projects often impact multiple business processes and enabling IT systems. To compound the complexity, most organizations have to contend with multiple projects impacting the enterprise at any given time. The modern decentralized enterprise thus requires us to better manage interdependent change portfolios and to reconsider current project-centric management practices.

Enterprise release management (ERM) takes a holistic view of a change portfolio and its impacts on the enterprise. It combines traditional management perspectives including project/program management and enterprise architecture, and integrates them into a release-centric approach that borrows from software development's agile methods with its emphasis on iterative delivery and configuration management. The ultimate goal is to achieve a unifying management perspective of strategic change delivery at the appropriate level of abstraction. ERM's focus on the big picture offers business and IT leaders the tools to finally chart and track the business' strategic course without being distracted by the details that current management practices address.

This book provides guidance on the processes, tools, and techniques necessary to achieve visibility and governance of an enterprise's strategic evolution. It stimulates readers to rethink the way that organizations manage business transformation and deliver strategic change, demonstrating why the project as a management construct is less effective in the modern business context. Insights into what makes project success elusive provide the impetus for new portfolio-based release management techniques that enable a holistic perspective of an enterprise's change initiatives. ERM provides the tools to manage the portfolio life cycle and coordinate the attendant dependencies, negotiations, and compromises that have to be made by enterprise stakeholders.

A unifying release-centric paradigm is shown to be necessary for the planning and execution of strategy in the enterprise context. The case for ERM is made clear in organizations where a change portfolio comprising multiple programs of work impacts the enterprise architecture. In such environments, an integrated enterprise release consolidating the deliverables from multiple projects becomes the appropriate vehicle for delivering change—not projects! While projects remain the means of managing scope and organizing its execution, the entire portfolio needs to be integrated, tested, and released as a whole.

While enterprises have endeavoured to become agile by embracing change and aligning IT with business, this can be ineffective when attempted in isolation. The reality of shared services and IT infrastructure give rise to greater management complexity that is rarely addressed in planning and coordinating change programs. By embracing this complexity at the appropriate level of abstraction, this book provides simple tools that enable enterprise resources to be better deployed to optimize the delivery of strategic change. ERM can offer a business new oversight and processes to better track delivery progress while providing the tools to support management agility in delivering a strategic change portfolio.

1

The Evolving Enterprise

If a man will begin with certainties, he shall end in doubts; but if he will be content to begin with doubts he shall end in certainties.
—Francis Bacon

The term enterprise release management (ERM) needs both an introduction and some measure of justification. The word enterprise is overused; placed in front of any management discipline to lend it substance it can serve to lose significance. In this book, the term enterprise is used to describe a high-level integrating perspective of a multifaceted, distributed organization.

Achieving an enterprise perspective is notoriously difficult to attain, as it requires piecing together and synthesizing the different, siloed views of the myriad business and functional units across the organization. Traditional reductionist thinking has been good for breaking down organizational complexity through analysis and decomposition, but leaves us with the challenge of synthesis and integration. By introducing a generalized release paradigm, this book offers a means of achieving a consolidating, strategic perspective of an enterprise's evolution.

ERM is therefore about integrating and scaling a number of familiar development management disciplines to the enterprise level. In doing so, new patterns can be recognized and abstractions exploited to provide the management visibility needed to guide the delivery of a strategic change portfolio. It is this portfolio that directs the enterprise's evolution—and it is evolution. While the term enterprise transformation is increasingly popular, seeking to use such catchy phrases can overlook the need for operational continuity. As used in this book, evolution encompasses changes of different scale and granularity. This includes strategic transformations, since rarely does a transformation create an entirely separate and independent entity that does not have some operational dependence on the enterprise.

In setting out to explore ERM, this chapter first provides the motivation for an enterprise-level release. The different interpretations of release management are introduced and discussed before enlisting concepts from enterprise architecture to establish the foundation terminology for discussing enterprise releases.

1.1 Enterprise 2.0

The recently coined term Enterprise 2.0 [1] can serve as an example of both of the common use the word enterprise and to present an illustration of ERM in action. Intending to describe the adoption of Web 2.0 social media software like Twitter and Facebook to connect individuals and teams in the organization, Enterprise 2.0 is shorthand for "a system of web-based technologies that provide rapid and agile collaboration, information sharing, emergence and integration capabilities in the extended enterprise."

Enterprise 2.0 is therefore another example of organizations seeking to adopt new technologies and appropriate them to improve business efficiency. It represents the latest attempt to utilize the IT innovations to coalesce large, distributed organizations and provides a convenient metaphor for the evolution of an enterprise. Indicating another stage in the transformation of the enterprise, the number 2.0 represents a new version or release of the enterprise; a snapshot of the people, process, and technology dimensions of a continually evolving enterprise. The form of Enterprise 2.0 implies the existence of an Enterprise 1.0 that relies on more traditional enterprise systems without the benefit of social media capabilities. It also suggests that we can anticipate an Enterprise 3.0, or possibly a more incremental change to Enterprise 2.1; new releases of the enterprise that will come into existence as business opportunities are embraced or new processes and technologies are adopted.

Without considering the details of what new capabilities might be on offer for future incarnations of the enterprise, the simple example illustrated in Figure 1.1 demonstrates what this book sets out to achieve with the introduction of ERM: strategically managing the evolution of the enterprise using tried and tested approaches suitably scaled for the enterprise. Looking suspiciously like a product roadmap or a software version tree, the figure may be considered a manifestation of what tool vendors have termed enterprise life cycle management. But this natural extension of the Enterprise 2.0 metaphor illustrates ERM's goal of abstracting familiar engineering management disciplines and adapting the experiences of software development to strategy planning and execution. The relationship between software development and the

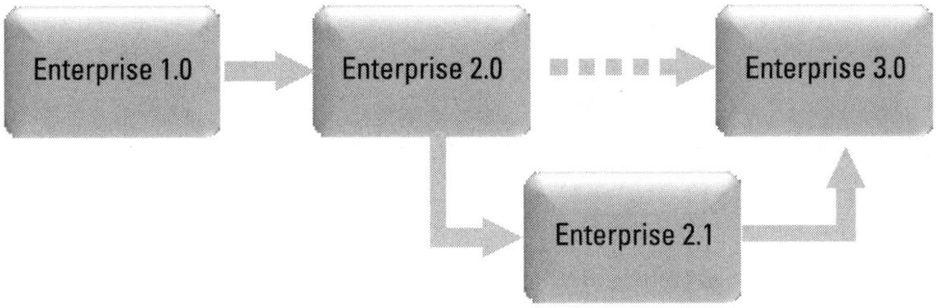

Figure 1.1 Enterprise releases.

implementation strategy should not come as too great a surprise. Treating the enterprise as a complex system or product presents the opportunity to consider and manage it holistically. In many ways, what executives do to determine and execute strategy corresponds to the role a product manager performs when identifying and prioritizing customer requirements. While there are of course crucial differences in scope, scale, and context, the decision making involved shares common characteristics. Cross-fertilization between these two fields can offer new management insights towards achieving a more responsive, agile approach to enterprise planning. Accordingly, ERM views the enterprise release as a generalization of a software release, and provides a framework for strategic execution based around a release paradigm.

A fresh perspective on delivering strategic change is warranted and timely. The limitations of current management best practice become more evident as the complexity of continually evolving enterprises increases. What the simple illustration (Figure 1.1) hides is the inefficiency and business risk involved in transitioning the enterprise from one state to the next. In the face of relentless change and growing business complexity, new paradigms are needed to achieve a holistic management perspective of an enterprise that can ensure visibility and stability of its evolution. To do that we need to balance the twin forces of change and complexity to successfully deliver business value.

1.1.1 Relentless Business Change

Change represents both an opportunity and a threat to modern businesses. In today's globalized economy that demands innovation, the management challenge is to be responsive to, and indeed encourage, change. Management needs to ensure that innovation is not stifled by excessively bureaucratic change governance. The business imperative to deliver new competitive products and services to market in as short a time as possible commonly results in multiple initiatives being undertaken in parallel. These concurrent initiatives form a portfolio of changes, often sponsored by different business units across the enterprise.

The management challenges faced under these conditions are unlike anything we have had to deal with in the past. Projects designed to ensure management focus can too often be blinkers instead, delivering changes that splinter the enterprise and fragment its architecture. Experienced practitioners understand the consequences of the incessant stream of changes that are encountered in the real world and the following quote [2] aptly describes the situation:

> In an effort to take advantage of valuable new opportunities, multi-project organizations, more often than not, tend to launch projects as soon as they are understood, concurrently with existing projects, simultaneously with other new efforts, and unfortunately too often, without sufficient regard to the capacity of the organization. A common result is that the responsibility for sorting out an array of conflicting priorities often falls to project resources and their managers ... the resultant locally set priorities may not be in synch with each other or, more importantly, with the global priorities of the larger organization.

Enterprises have spent the last few decades growing businesses, integrating markets and value chains, and consolidating the processes and technology infrastructure. They provide customers with comprehensive services and products, such as one-stop-shopping for financial services and integrated smart devices. Internally, these organizations have been breaking down the functional silos, replacing them with flat organizational structures and federated business models supported by increasingly shared services. And all the while we have been accruing a management debt; the burden of managing highly interrelated and interdependent business and technology systems. The modern enterprise has grown to become a distributed and dynamic organization; a complex adaptive system. But, as yet, we have not upgraded our management practices to suit these new distributed and federated business models.

1.1.2 A Necessary Management Synthesis

Even as businesses continue to grow increasingly sophisticated, the management tools and techniques used to deliver Enterprise 2.x, whether it is based on new products developed for customers or IT capabilities supporting business processes, are showing the strain. In these competitive times, the inability to deliver strategic new capabilities in a timely and cost-effective manner can be devastating to the enterprise. Yet we have persistent evidence that only a small number of projects deliver the expected benefits. The Standish Group's Chaos Reports have, over the years, consistently been the harbinger of the woeful state of project execution, with the probability of successful delivery stubbornly remaining around one-third of all IT projects initiated [3]. While researchers have challenged the Chaos Report for its lack of transparency and rigorous methodology [4] and the apparent paradox of the high rate of IT adoption even as projects appear to routinely fail [5], these results touch on a shared perception across the industry that projects are generally challenging—and the risk of failure grows with the size and volitality of a project [6].

A sense of dejected resignation can be detected when talking with experienced program and delivery managers today; a growing suspicion that many of the management practices that we hold dear and teach to new practitioners are often inadequate in real-world situations. There can be a gulf between the guidance and advice offered by best practices versus the reality that practitioners face in the trenches of the modern business. That gulf is discomforting at a time when some of our trusted techniques are approaching their fiftieth anniversary. At the time of writing this, the critical path method (CPM) a classic PM discipline, already has that distinction. Of course, any gaps in our body of knowledge are usually filled by years of experience, but that tacit knowledge is difficult to teach and essentially misses the point of having a documented and codified management framework. The introduction of agile methods (to be discussed in Chapter 2) has been seen to topple some of the original thinking on projects and management control. But the challenge to the existing orthodoxy is presently confined to dedicated software development projects and arguably does not go far enough.

Now is the appropriate time for an appraisal of what it is we practice in the name of management. And address the difficult questions that challenge management to be more engaged and responsive, rather than simply being an expensive overhead that to some represents a substitute for poor collaboration and communication. The new realities in the enterprise call for new management innovations and novel approaches. The release-centric techniques that this book introduces are not heretical concepts, rather they represent a necessary synthesis to ensure our management arsenal keeps pace with Enterprise 2.0. Unlike management fads, ERM will be shown to be a natural extension to today's practice, founded on the bedrock of management science, systems engineering, and program management.

1.2 Unifying Release Management

The concept of a release, used widely in product development and management, can be generalized to be a unifying paradigm across the different disciplines needed to plan and execute strategy. The fact that the word release is commonly applied as both a verb and a noun provides evidence for both its ubiquity and its potential for establishing a shared vision between business and

IT. A supplier may release a new version of a product while the customer has to implement that release; in the first instance release is used as a verb and in the second it is a noun. So the word release is clearly rich with meaning and can be interpreted either from a product or a process perspective.

As a result, current release management terminology can be confusing and overloaded, having multiple connotations that corrupt its meaning and reduce the influence of this nascent discipline. Like many terms used in the volatile world of IT and software development, it can mean different things to different people. Several of the common yet somewhat conflicting meanings will be reviewed here so readers can identify their own inclinations and biases, but appreciate alternative viewpoints before a more comprehensive and generalized perspective of the release is achieved.

ERM attempts to clarify and extend the concepts of release management, applying them to the delivery of strategy in the enterprise. As an evolving, complex adaptive system (or more appropriately in the modern, devolved enterprise, a system of systems) an enterprise release can focus and align current business and IT management practices. To gain the benefits of ERM, however, requires a more integrated, unified approach to managing the end-to-end release life cycle, from the inception of strategic initiatives to the execution and delivery of business benefits. While some might argue that this reframing of release management could introduce yet another set of methods to an already crowded space, it should be made clear that the generalization undertaken in Chapter 7 actually provides a rationalization of common usage that yields a powerful management abstraction based on the release paradigm.

1.2.1 Historical Context of the Release

In lay terms, a release is defined as the setting free from restraint. In solution delivery it is the culmination of what is often a heroic team effort; the end result of planning and development processes that finally yield a tangible product or service. A release is universally recognized as a critical milestone and its historical context serves to remind us both of its pedigree and the integral role it can play through the development life cycle. To appreciate its significance does not require the detailed understanding of its genesis, but it does demonstrate why release management is emerging as a powerful discipline in its own right.

The many lenses through which a release can be considered are both a representation of the power of the paradigm and the difficulties of capturing the essence of the release in a manner that is relevant and appropriate to the myriad stakeholders across the product life cycle. Common usage of the term and its increasing acceptance indicates however, that the release, as a concept, has the potential to integrate these different perspectives.

1.2.1.1 The Origins of the Release

In the not-too-distant past, the final stages of project or program delivery comprised a series of one-off activities necessary to deploy or set-to-work the product configuration. The importance of these latter stages of delivery were recognized and enshrined in early Department of Defense (DOD) and industry standards that focused on the configuration management (CM) discipline and its emphasis on the integrity of the product configuration as it was released. The reviews and audits that CM standards specified, in conjunction with related project management, systems engineering, and quality assurance disciplines, defined in detail the steps for releasing the complex engineering systems that were being developed at the time.

The early history of CM is intriguing, with the discipline being hailed for bringing governance to defense programs during the Cold War era [7]. However, as an engineering management discipline originally intended to manage mechanical and electrical components (an early mantra of CM being "form, fit, and function" of components), it was seen as a technical and administrative necessity and was rarely accepted as a mainstream management discipline. The detailed focus CM placed on maintaining registers, release records, status reports, reviews, sign-offs, and audits kept it in the administrative domain; essentially a coordinating discipline that assisted those with the technical management responsibility, or authority, to make the decisions needed. The real power resided with the program or project managers (henceforth denoted simply as project managers except where the distinction is relevant to the discussion) who took the information CM provided and liaised with customers and sponsors to ensure their expectations were being managed.

Focusing on the product perspective, the CM discipline ensured that the developing configuration was in a known state, with clearly identified baselines recording its emerging specification and the status of any changes. These baselines, aligned with project milestones, with CM activities ensuring the integrity and consistency of the product concept at these checkpoints in the development life cycle. In this way CM ensured the product perspective was intertwined with the development process and supported the quality reviews and approvals necessary before the next phase of development could commence.

CM's less-than-glamorous role had to deal with the engineering reality and account for the status of the product development. While the formality surrounding CM could become onerous, the ultimate goal of the governance and assurance processes it enshrined was to ensure the release of an acceptable product. Where CM focused on controlling the integrity of the product, a release comprised those activities, or processes, that delivered the product to the next recipient in the development life cycle. The link with formal CM can be seen in the following definition of a release as [8]:

> …a configuration management action whereby a particular version of software is made available for a specific purpose (e.g., released for test).

More recently, these release processes have come into greater prominence not simply because of the sheer volume of projects being undertaken, but due to the advent of software technologies that make releasing new capabilities a routine aspect of development. This caused a significant shift in the focus of the CM discipline as it took root in software development—not least in the separation of the CM discipline into two constituent parts of change and configuration management. While the changing emphasis of the CM discipline is not the purpose of this book, the fact is that these apparently established disciplines continue to evolve, providing new interpretations of the release and the nascent release management discipline. And the good news is that release management become a much less onerous process that can be effectively integrated into the delivery life cycle.

1.2.2 Release Management in Software and IT

Adoption of CM in the IT domain was initially slow and problematic given the level of rigor demanded by the discipline. A product or system configuration needed to be suitably complex to warrant the effort and costs associated with CM. Yet, key aspects of the CM discipline have crossover into the mainstream, are widely practiced in product engineering, and have had tremendous growth in software development where the advent of software CM tools have removed much of the administrative drudgery associated with the discipline. With these tools came increasing automation, reliability, and frequency of building and releasing software applications that have seen greater interest in the more dynamic (as in the automation of CM processes) release management discipline and related development capabilities it can enable, such as continuous integration [9].

A separate impetus behind release management is the recognition of CM's role in managing IT operations. This natural evolution of traditional hardware CM as it applies to IT infrastructure is evidenced by the growing interest in the IT Infrastructure Library (ITIL) [10] and the availability of software tools termed, somewhat generically, configuration management databases (CMDBs), which aim to capture the configuration of the networks, computing assets, and applications that underpin IT operations.

As we gain the abilities to better manage the configuration of the various "product domains" that play across the enterprise, addressing the need to coordinate changes and releases across these different (potential) silos represent the next management frontier. In taking these steps, the release is becoming *the* configuration itself—a noun—the entity being managed as well as the process for delivering it to the world. Release management has become a modern variant of CM, albeit placing an emphasis on final stages of product or solution delivery—not surprisingly given the ease with which a software product can be released—with its scope often extended to the distribution of the software packages to different computers on a network. As evidence, note the shift from a process to a product emphasis in the following definition of a release provided from an IT services standpoint in the ITIL framework:

> A collection of hardware, software, documentation, processes or other components required to implement one or more approved changes to IT services. The contents of each release are managed, tested, and deployed as a single entity.

The richness of the release concept across the enterprise is at the heart of the ERM that taps the concept's emerging relevance to management. There has been little recognition of the converging trends that see a release as the common thread running through strategic planning, execution, delivery, and operations—each of which aim to manage the impact IT has on the enterprise and impose governance over its evolution.

As the release construct and release management becomes more pertinent in the modern enterprise, it can be seen to subsume older disciplines like systems engineering and CM that gave birth to it. Perhaps this is due to the difficulty CM had in gaining traction as a management discipline or the awkwardness of its configuration-centric terminology. Perhaps it is due to the influence of software development tools that enable rapid, iterative releases and continuous integration. Whatever the reason, the release is a more visible management entity that is at the cutting edge of innovation in both software development and IT management. By synthesizing earlier disciplines like CM, the release is poised to become a unifying paradigm for enterprise transformation and change management.

1.2.1.2 Perspectives of the Release

The historic overview of the release shows it has undergone a metamorphosis in recent times, with different groups having their own interpretation and understanding of the concept. It is the goal of the ERM to attempt a unification of these viewpoints and to assemble a multifaceted understanding of the release from the point it is first identified and defined as a part of strategic planning to its ultimate realization as a solution, be it a product or a service, which delivers the promised benefits.

As shown in Figure 1.2, the release can be the nexus between process and product views of solution delivery; the touchstone bringing together what needs to be delivered with the critical activities that make that possible. A release is about realization; it is where the rubber hits the road, or more accurately,

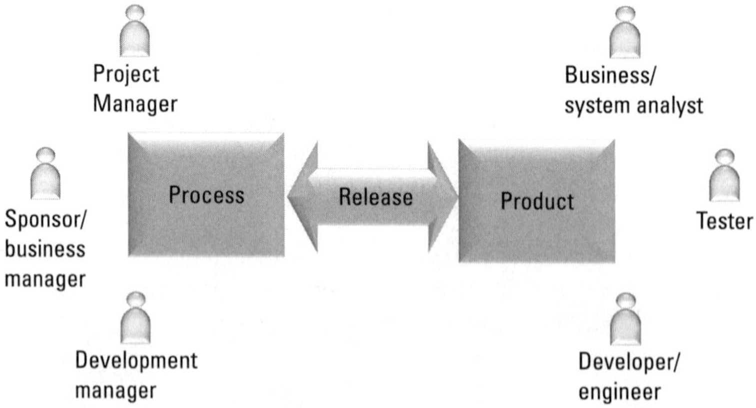

Figure 1.2 The release as process-product nexus.

where the process delivers the product. The connection that the release provides between process and product is a consequence of its emphasis is on the end-game. At the pointy end of a project, when a milestone has to be delivered or an outcome achieved, it becomes immaterial whether one is focused on the product or the process; the two converge and the release simply has to happen.

The release's ability to be interpreted from either a process or product perspective helps to bring together the different stakeholders across the solution delivery life cycle. Sponsors and managers of an initiative are likely to take a high-level, more abstracted process-centric perspective of the initiative that considers the activities needed to delivery the business outcomes. It is left to the specialists in IT, engineering, and other functional areas to focus on the details of the product specification, development, and testing. And it is the release that acts as the lingua franca between these different roles and the separate viewpoints they take of the initiative.

A process focus takes an abstracted life cycle view of the release that can serve to orchestrate the different specialist activities so they flow towards the ultimate delivery of the solution. For the action-oriented project managers, a release represents a future milestone to be delivered with the processes steps necessary to achieve that end. Their concern is to navigate the solution delivery life cycle and manage the budget and schedule dimension of the triple constraint. From a product perspective, the focus is on the scope that is ultimately delivered in the release. This is where the contributions of the product team(s) involved across the life cycle finally gain visibility and are subjected to the harsh reality of testing and user acceptance. This is the view traditionally taken in disciplines like CM, where the release (used here as a noun) is the product configuration that is delivered. But the inflection point for release management, the point where it moved closer to a process orientation, arose when the release ceased to be that one-off delivery of a new product. It instead became a more regular activity, requiring its own processes, associated with the continual incremental delivery of an evolving system or product—and in the ERM context, the enterprise.

The need to repeatedly coordinate the delivery activities undertaken by a multidisciplinary team and ensure an integrated, coherent delivery accentuated the process dimension of the release. This has had the consequence of most common interpretations of release management being confined to the latter, delivery stages of the product life cycle—even encompassing the packaging and distribution of a software application over a network. Yet the act of releasing was never a singular event in the product life cycle and while the ultimate recipient is the end user or customer, there would necessarily be a number of intermediate stages, including integration, test, and acceptance through which a release must first pass. Those who view the release as a process, therefore, see a series of states through which a release must transition and that together define a more comprehensive, end-to-end release process or life cycle. As the release goes through these states, the successive release milestones become shared goals that bring together the different perspectives of managers, developers, integrators, testers, and end users.

The release has taken a more prominent role in IT management with the adoption of iterative development methods that advocate multiple incremental releases of the required capabilities. Shorter project delivery cycles see development projects planned as several releases, each providing a part of the desired capability. In such cases, there is no longer a single delivery; rather the customer can be actively engaged in reviewing, testing, and even using the capabilities delivered in a series of release milestones. As the number of releases grew, it ceased to be a milestone and instead took on more a process focus. The interpretation of a release as a process that guides the development of a work-in-progress, evolving product is growing in acceptance and creating a new awareness of the release as a relevant management construct across the entire product life cycle.

The original focus of a release on the sharp end of delivery has been changing with greater emphasis on the early stages of a release life cycle; although it is the significance of the delivery milestone that provides the impetus to understand how the release comes into being. Instead of simply managing the technical integrity of the final solution, release management increasingly encompasses earlier phases such as release planning, where requirements have to be prioritized and assigned to current or future, planned releases. This emphasis on the early phases of the life cycle can be attributed to shorter delivery cycles and increased collaboration needed between the different disciplines involved in the release. It became easier for both users and developers to see the consequences of poor upstream decision making and the less formal, contractual relationship between users and software developers allowed greater understanding of the importance of clear requirements and effective release planning.

Therefore, the interpretation of the release to be taken henceforth has a broader relevance beyond the latter, delivery phases of the product life cycle to encompass an end-to-end perspective that starts with the very inception of a release. This wider view and process interpretation of the release is in many ways synonymous with the product development process or IT's solution development life cycle (SDLC)—the distinction will be shown to lie in the ascendancy of the release resulting when a portfolio of changes has to be managed.

Beyond the release life cycle already evident, further generalization of the definition of a release is needed if we are to achieve enterprise releases as exemplified in the expression Enterprise 2.0. To achieve this, current interpretations of release management have to be merged, extended, and scaled to the enterprise level. Enterprise releases need to be recognized to address the many dimensions of the business that have to be managed and aligned. A generalization of the release management discipline is needed to manage and guide the execution of enterprise strategy. And a new interpretation of a release is essential, one that is not confined to the technology dimension alone. Instead, the entire enterprise must be considered as a configuration and release management extended to address business impacts so that the evolution of the enterprise can be managed holistically.

1.3 The Enterprise Perspective

Achieving an enterprise perspective is essential for informed decision-making across the range of business activities undertaken in an organization. Having a view of all the factors effecting a decision can assist business leaders to make astute, expeditious decisions that weigh strategic and operational priorities and make the appropriate trade-offs. Whether an organization's strategy is to grow its market share, create innovative new offerings, or reduce costs, considering the complete enterprise picture can pay dividends in the selection and delivery of a balanced change portfolio.

Taking such an enterprise perspective is fundamental to ERM as it attempts to describe, and indeed guide, the evolution of an organization. Larger and more complex enterprises the comprising multiple lines of business, benefit from the language developed in the field of enterprise architecture (EA), which provides an abstract view of a company, organization, or enterprise [11]. EA is the latest effort that aims to model the enterprise; essentially deconstructing a business in an attempt to understand its constituent parts and the interactions between them.

Aside from, and possibly because of these grand ambitions, the challenge many enterprise architects face is to better engage in strategic business deliberations. While there is a perception that EA can be an idealistic, ivory-tower discipline, it offers a framework that helps in the discussion and elucidation of ERM. Further, ERM itself provides a means of clarifying the critical contribution architects can make in the planning and execution of enterprise strategy, ensuring their deliberations actionable and relevant. While a detailed understanding of EA is not a prerequisite for ERM, it provides the appropriate context for describing enterprise change and release and frames the ERM narrative.

1.3.1 Overview of Enterprise Architecture

Growing business reliance on IT and the increasing sophistication of enterprise systems has been the impetus for changes to be understood and governed from an enterprise perspective. EA provides a framework for planning the future or target state of the enterprise and aligning changes in business and technology with enterprise strategies and priorities.

The Open Group, publishers of the widely adopted TOGAF framework, defines an enterprise as "any collection of organizations that has a common set of goals and/or a single bottom line." It gives the primary reason for developing an enterprise architecture as aiming to "achieve the right balance between IT efficiency and business innovation." TOGAF goes on to state that a good enterprise architecture allows individual business units to innovate safely in pursuit of competitive advantage while assuring an integrated IT strategy that permits synergy across the extended enterprise.

1.3.2 Modeling the Enterprise

EA can be considered an example of a reductionist response to managing complexity and applies decomposition to better understand the enterprise's components and their relationships, both present and future. The EA discipline attempts to model the enterprise by considering it from five different perspectives, which should all align in a coherent enterprise. Different EA frameworks place different emphases on these dimensions but all support strategic execution by developing plans and blueprints that guide decision making about enterprise change and transformation. The key dimensions that EA uses to represent an organization are:

- The strategic dimension that attempts to capture the strategic goals and initiatives;

- The business dimension that looks at the products and services marketed by the organization as well as the business structure and processes that are used to develop and support them;
- The information dimensions that consider the data needed to operate and manage the business, the data flows needed to complete transactions, and the processes that control the integrity and security of this data;
- The application dimension that considers the information systems and software products or packages that support the automation of processes and data flows;
- The infrastructure dimension captures the network topology, telecommunications, and facilities that are essential to support the organization.

Bernard's EA cube, illustrated in Figure 1.3, is indicative of EA framework and highlights the enterprise architect's aspiration to have the architectural dimensions align and be managed consistently across different lines of business. Such a holistic enterprise perspective aims to achieve greater standardization across the enterprise that drives greater operational efficiencies. While in reality there will be differences arising from departments or entire businesses being at different stages of maturity and effectiveness, the exceptions to enterprise standards and practices can then be the focus of management attention that aim to bring them into alignment.

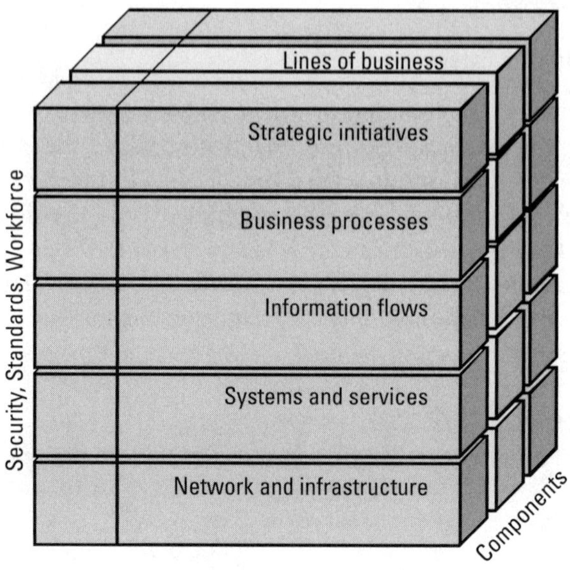

Figure 1.3 The enterprise architecture cube. (*Adapted from:* [10].)

While enterprise strategy can be seen to impact all dimensions, it is convenient to split the management of these dimensions into the executive, business, and IT responsibilities. Adhering to this convention, in what follows, the executive dimension is responsible for strategy and governance, the business and information dimensions will be discussed as part of the business function (and often grouped together as the business architecture), while the application and infrastructure dimensions considered the domain of IT (described as the technology architecture). In more colloquial language, an enterprise's architecture can be simply considered to be the people, process, and technology dimensions of an organization that is managed at the strategic, tactical, and operational level.

Most enterprises have evolved organically and so do not easily fit into the neat conventions and orderly architectural dimensions prescribed by EA. For example, it is common for IT to assist a business in maintaining its business processes even though it is strictly speaking their responsibility. Similarly, it is a convenient simplification to consider that a business alone determines its enterprise strategy with IT's role relegated to implementing the system changes needed.

The reality is often more complex and what creates an impetus for strategic change can be rich and varied, with disruptive technologies like the Internet and the World Wide Web showing how business-driven strategies can at times be turned on its head. Yet, in recognition of business as the revenue generator and the ultimate financial sponsor of IT investments (and to keep things simple), the evolution of the enterprise will be described from a business-driven strategic perspective.

1.3.1.1 Enterprise Strategy and Planning

If EA aims to guide enterprise evolution and provides the necessary mechanics to address enterprise change and transformation, it can sometimes work to a different timescale to that of the business. The discipline offers a methodological basis for designing and specifying the different architectural dimensions needed to help the organization understand, oversee, and realize the enterprise strategy and planning its eventual execution—even though, in reality, it is not always possible to follow it religiously. While EA provides the means to describe and plan the evolution of complex, distributed organizations, the pace of business change can demand greater agility. There is a clear benefit in having EA be more closely aligned with the faster-paced, delivery-focused disciplines like project and program management.

Without attempting to specify the manner by which an enterprise's business strategy is determined, EA traditionally aims to provide the blueprint for enterprise evolution [11]. This idealized view is abstracted from the details of the business itself even as the ultimate end-product of strategic planning, and the result of any transformation program, is one or more enterprise releases—essentially new versions of the enterprise, evolved to better adapt to the business conditions and executing the strategic vision.

Figure 1.4 illustrates the common terminology employed by enterprise architects to describe the transitional states of enterprise evolution. The current and target states of the enterprise are more informally and affectionately known as the AS-IS and the TO-BE states. The current state represents the existing strategy, business, and technology architectures guiding and supporting the present business operating model, sometimes referred to as operations or business-as-usual. The target state represents the next transitional step towards implementing the ultimate blueprint for the enterprise (or some subset thereof). This evolutionary picture corresponds nicely with the Enterprise 2.0 metaphor. Each target state represents an enterprise release; a stable and transitional architectural baseline on the path to the ultimate future state of the organization envisioned in the strategic plan.

EA provides the basis for representing the enterprise's target states (such as Enterprise 3.0 and subsequent incarnations) in terms of the potential impact that changes will have on the enterprise's architecture—much as product management establishes a release roadmap for a product and its architecture. Strategy execution can therefore be viewed through the prism of the four enterprise dimensions. Each dimension requires its own architectural blueprint for the target state, which must align with each other to ensure the enterprise evolves in an orderly fashion to execute the agreed strategy.

While enterprise architects may actively participate in describing and assessing the current state and deriving and documenting the future target states, they are less involved in planning the transition between the states and at best play a design governance role in the execution of the plan itself. Instead, the changes required to transition the enterprise to the desired target state become part of the change portfolio and are commonly delivered through one or more projects, or a more comprehensive transformation program. While EA attempts to provide a framework for planning a strategic course for the enterprise, the

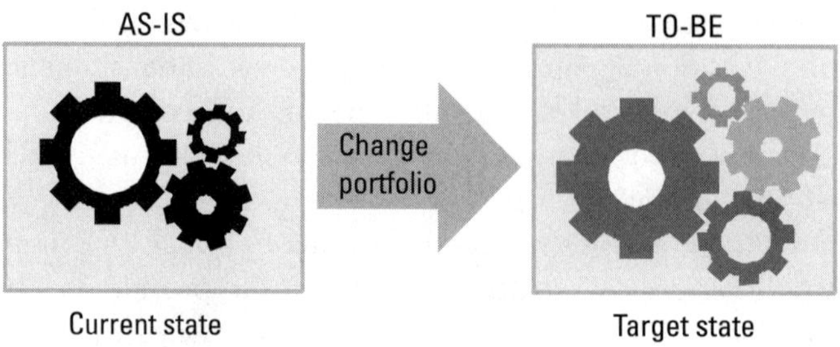

Figure 1.4 Current versus target state.

responsibility for the delivery of the target-operating-model is generally handed over to the program and project managers for execution.

Therefore, while EA has developed a rich language to describe strategic planning at an enterprise level that aligns business and IT, it remains for executives and business leaders to establish the appropriate strategy for the enterprise and work closely with other governance functions to execute the transition effectively. In reality, the development of a strategy plan for the enterprise is less than scientific and more akin to a black art, with many executives not appreciating the rigors of a defined planning process. Put simply, senior executives do not like to be constrained in the development of strategy and prefer to trust their business instincts. Yet when it comes to executing strategy, there are inherent constraints that require a balance be struck between business innovation and the effective management of the platforms that support the business. So, while the development of the strategy may be the preserve of business executives, it is in planning the execution of that strategy where the role of the business and program managers introduce pragmatic considerations that can limit the range of viable strategic options.

Where EA takes a deterministic approach to enterprise strategy, it can become sidelined as most enterprises are volatile environments that harbor more than a trace of uncertainty and randomness. So, while the EA offers a solid basis for the discussion enterprise releases and ERM processes, it needs to be closely integrated with the planning to be relevant in chaotic business environments where the determination and execution of strategy can be more spontaneous and dynamic. In such environments, it is rare for an enterprise strategy to be the sole driver of the change portfolio. Allowances need to be made for practical adjustments to the transition plan as well as accommodate ad hoc, unplanned initiatives resulting from external factors such as legislative changes or regulatory compliance.

1.3.1.2 Agile Enterprise Releases

If everything in an enterprise worked predictably and in accordance with an approved strategic plan over an extended timescale there would be no need to consider more agile planning cycles. The traditional, annual planning cycle would suffice and there would be little need to consider how an enterprise strategy was developed and executed. Like early systems development, the definition of the to-be state of Enterprise 2.x would be sufficiently stable, the pace of transition and execution so slow, that there would be little need to establish an enterprise release process. The delivery of a new enterprise release would be so infrequent that the enterprise could muddle through with the concerted and dedicated effort of individuals in key roles without too many problems.

Instead, as we discuss in the next chapter, we live in a world of rapid change where enterprise strategies need to adjust to business conditions. Leadership must be responsive, closely monitoring strategic initiatives and adjusting course as business priorities shift. Strategy development and execution need to become increasingly flexible, adaptive and, ironically, short-term focused. The requirement for correspondingly more adroit and agile governance processes in the enterprise now begin to mirror the changing attitudes to management in software development. The need to focus more closely on enterprise evolution and the transition states, or releases, necessary to execute strategy become more pronounced.

While the delivery of business strategy is not usually couched in IT derived terminology such as a release, there is nonetheless a clear analogy between trends in software or product development and the increasing pace of enterprise evolution and transformation. As a result of the need for greater business agility, there is common ground to be found with the agile methods that have been a powerful influence in software development. In treating the enterprise as a complex system, or system of systems, the ERM leverages the experiences gained in managing continually evolving software systems and applies them in the enterprise context. The necessary scaling of these concepts to the realm of strategy planning and execution cannot, however, be undertaken lightly. While the analogy may be compelling, adjustments are to be expected and any flaws that are evident "in the small" will be magnified and potentially catastrophic at the enterprise-level. Hence, new techniques will need to be identified, understood, and explored to address weaknesses in our existing interpretation of release management so it is robust enough to manage enterprise releases.

The challenges of an evolving enterprise operating in a dynamic, market-driven environment is the motivation for ERM. As the implications of an end-to-end, life cycle perspective of the release are detailed in this book, the power of taking what we term a release-centric focus should become clear. This sees the release central to strategy development, planning, and execution with ERM providing the practices, techniques, and tools to support more responsive governance that is better suited to today's volatile business environment.

The positioning of ERM is literally presented in Figure 1.5, placed centrally between the twin challenges of business complexity and relentless change. The ERM is shown as a neutral approach that can moderate the competing management philosophies of command and control versus collaboration and agility. The common characteristics of these two approaches to solution delivery book-end ERM in the diagram and will be expanded upon in the next chapter. The expansion of ERM in the central panel attempts to show it as a scalable management framework that can support different life cycle models. The focus on a release offers a more general representation of the solution delivery that has been termed the release life cycle and that offers improved management visibility and reporting.

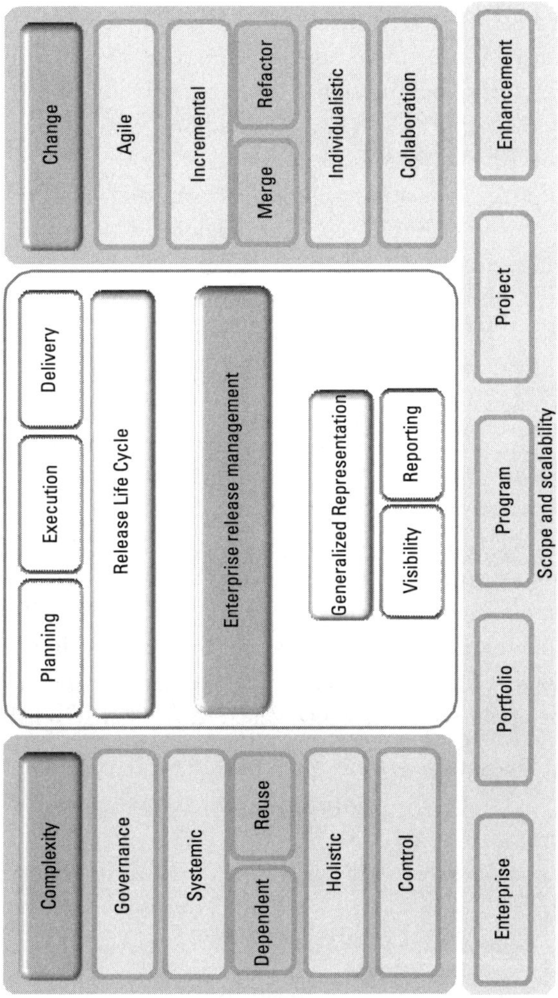

Figure 1.5 Features of ERM.

The above of course represents the vision for ERM that is to be fleshed out in subsequent chapters. As a comprehensive reengineering of traditional solution delivery, it can encompass the planning, execution, and delivery of strategic initiatives—which increasingly have to be integrated into an enterprise release. There is growing evidence that the release-centric perspective is gaining traction with the role of senior release and delivery managers becoming more common. But where these roles reside in the enterprise and what tasks they have to perform varies greatly.

There is an evident need for a scalable framework for release management, even where that may remain aspirational. ERM adoption can be expected to be initially confined to specific groups and focussed on troublesome phases of the life cycle. Aspects of ERM already exist in more mature organizations today, but the full benefit of a release-centric approach to strategic delivery will emerge only when the enterprise begins to "connect the dots." That requires the release to be recognized as a simple and elegant paradigm that can connect all parties across the life cycle responsible for delivering strategic change. The remainder of this book sets out to advance that argument, presenting both a justification and an exposition of ERM; showing how it brings together many familiar management practices in a new integrating framework that better address the management challenges in the enterprise. The next chapter therefore takes a step back to review the problem space and establishes why, in the face of a changing and increasingly complex business environment, we cannot simply rely on current management best practices to deliver a strategic change portfolio. It exposes the weakness in our management arsenal that makes us choose between agility and planning, as if just one must be deemed good and the other bad. Instead, an argument is presented for a higher-order union of these contending philosophies, one that can be achieved at the governance level that allows the most appropriate approach to be applied when executing a strategic change portfolio.

References

[1] Frappaolo, C., and D. Keldsen, *What is Web 2.0?*, Association for Information and Image Management, 2008.

[2] Patrick, F. S., *Program Management—Turning Many Projects into Few Priorities with TOC,* Presented at PMI Symposium, Philadelphia, 1999 (http://www.focusedperformance.com/articles/multipm.html, retrieved August, 2011)

[3] The Standish Group, *CHAOS Manifesto,* http://www.standishgroup.com/.

[4] Eveleens, J. L., and C. Verhoef, "The Rise and Fall of the Chaos Report Figures," *IEEE Computer Society,* Vol. 27, 2010, pp. 30–36.

[5] Glass, R., "The Standish Report: Does It Really Describe a Software Crisis?" *Communications of the ACM,* Vol. 49, 2006, pp. 15–16.

[6] Sauer, C., A. Gemino, and B. H. Heich, "The Impact of Size and Volitality on IT Project Performance," *Communications of the ACM,* Vol. 50, 2007.

[7] Johnson, S. B., *The Secret of Apollo: Systems Management in American and European Space Programs,* Baltimore, MD: John Hopkins University Press, 2002.

[8] Defense System Software Development, DOD-STD 2167A, 1988.

[9] Humble, J., and D. Farley, *Continuous Delivery: Reliable Software Releases through Build, Test, and Deployment Automation,* Reading, MA: Addison-Wesley Professional, 2010.

[10] ITIL V3. *Information Technology Infrastructure Library,* United Kingdom: Office of Government Commerce, 2010.

[11] Bernard, S. A., *An Introduction to Enterprise Architecture,* 2nd ed., AuthorHouse, 2005.

[12] Spewak, S. H., *Enterprise Architecture Planning: Developing a Blueprint for Data, Applications, and Technology,* New York: Wiley, 1993.

2

Embracing Change and Complexity

Our fast-paced and connected world grows increasingly integrated, interdependent, and complex by the day. In many ways the promise of digital technologies have been delivered and are becoming pervasive in modern business, as they have in our personal lives. We are all connected and instrumented as individuals, our status communicated and regularly updated—even if Enterprise 2.0 is not yet a reality in business.

IT has been instrumental in enabling successful businesses to expand by streamlining their processes and achieving a competitive edge in this fast-paced, global competition. Once-separate market segments, products, and indeed industries, have merged, converged, and coalesced. Modern enterprises now offer a dazzling choice of sophisticated products and integrated services designed to excite the marketplace and captivate potential customers.

Of course not everything is rosy; there is a price that has to be paid for this abundance of riches. The benefits of integration and amalgamation come with a management tax; the need to coordinate an increasing number of stakeholders, suppliers, and technologies that are increasingly intertwined and interdependent. There are new challenges for business in this fast-paced world where competition drives the adoption of the latest technologies and demands that we continually delight our customers with new offerings and capabilities. The great strides we have made must therefore be tempered by the ever-increasing complexity of the products, organizations, and supply chains.

Any enterprise, irrespective of its size, wishing to compete effectively today must adapt to changing business conditions and harness the increasingly complex business and IT architectures. Competitive advantage, and therefore business survival, relies on an organization's ability to select the appropriate strategy and translate it into the timely delivery of products and services to customers. Yet successful execution of enterprise strategy is problematic in a changing world and has to be achieved while continuing to support the current business model and the processes and technologies that are integral for day-to-day operations.

Today's successful enterprises need to balance the current state with their attempts to achieve the desired future state; there is a need to be agile and responsive so strategy can be adjusted and tweaked to accommodated changing market conditions. This challenge can pit those driving business changes against those responsible for delivery, such as IT departments and project managers, who must implement these changes while exercising responsibility for business and system continuity. Leadership requires better visibility of the challenges to executing strategy in order to craft realistic plans that can be adjusted to meet changing market conditions.

To explore the need for, and means of, attaining enterprise agility, we first consider the challenges in the enterprise in terms of change and complexity and study the interplay between these elemental forces. Finally, managing the delivery of enterprise change in the face of growing business complexity is shown to be an unsolved problem that is inadequately addressed by both agile and more traditional, plan-driven practices. Instead, successful delivery of enterprise change requires new management paradigms that consider both the business imperatives driving change and the inherent complexity of the enterprise architecture.

2.1 Wrestling with Change and Complexity

The use of IT in the enterprise has both made it amenable to rapid change as well as created a problem of growing complexity. The current state of most enterprise architectures is unlikely to make a CIO proud. Continual investment in successive generations of enterprise systems has created an accumulation of legacy systems that only serve to reduce the agility of the enterprise.

The challenges faced in delivering strategic change in the enterprise can therefore be framed in terms of how management responds to unrelenting change and growing enterprise complexity.

As shown in Figure 2.1, management in the enterprise can be considered to be constrained, if not defined, by these twin elements that frame the subsequent discussion on the adequacy of current management approaches. While the effect that these two fundamental and opposing forces are not simply confined to IT, the pivotal role that technology plays in enabling business process and information flows makes addressing them critically important. ERM aims to provide the processes and tools that help strike a balance between harnessing change and rationalizing increasingly complex business architectures; thus providing a framework for strategically managing enterprise evolution.

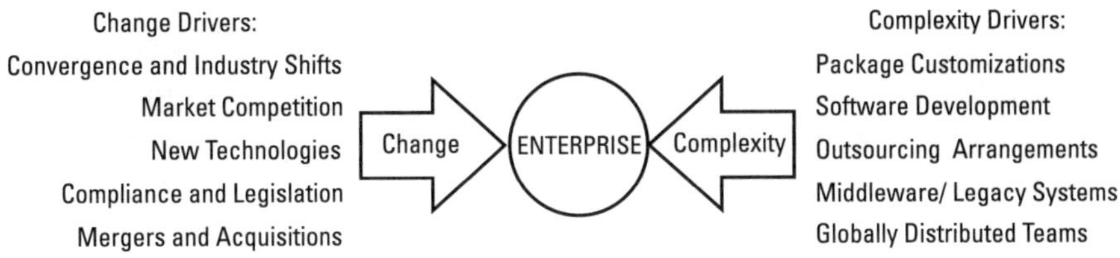

Figure 2.1 Change and complexity.

2.1.1 Relentless Business Change

Today's business environment is defined by change, whether it is simply changes brought on by organic growth, acquisitions, and mergers, or changes in the legislative environment. As a result of this churn, we live and modern businesses operate in what futurists have termed "permanent white water"—the milieu that makes enterprise changes a regular if not routine occurrence.

These changes have been propelled by the digitization of information and the resulting convergence of previously distinct products, and even industries. Moving to software-based, digital technologies has had immense impact on market expectations. IT solutions have increased business' ability to adapt and change to a point where reengineering is so common today we no longer even to use that expression. As processing power rises and the price-performance ratio tumbles, there is global competitive pressure for new or improved products and services. In today's flat, globalized world [1, 2], "time to market" has become the management mantra with continual waves of change sweeping over the enterprise as it looks to innovate and reinvent itself to be more competitive.

These fundamental, competitive forces have resulted in an insatiable demand for change, far greater than most organizations can deliver, and possibly absorb themselves. The constant flow of projects attempting to harness technology's potential give rise to a stream of multiple change projects, a common occurrence in most enterprise. Projects are now in fact so ubiquitous, Chapter 3 argues they can be considered business as usual and part of the enterprise landscape.

The Internet and the World Wide Web, combined with business processes automation, have led to the emergence of true e-businesses, even as the expression has become somewhat passé. In a relatively brief space of time, we have gone from databases providing electronic facsimiles of original paper-based customer records to sales executives being able to query and update enterprise information on their BlackBerrys or iPhones. Next, we can anticipate a revolution in enterprise communications and collaboration driven by the innovative use of social media technologies within the organization that take it closer to the tantalizing promise of Enterprise 2.0.

The malleability of software has driven innovation and the reality of continual change and striving to be first to market has not surprisingly changed management thinking. More than the sheer number of change programs in the enterprise, the pace of change meant that projects faced reprioritization and reassessment during the course of the project itself. Uncertain or changing requirements needed a new management style to be embraced that facilitated these changes and yet could be successful in delivering business outcomes.

2.1.1.1 Management and Agility

It is important to recognize that change at current levels has been a relatively recent phenomenon. It was less of a problem with earlier, hardware-oriented technologies where it was simply not possible to adjust requirements or plans without going back to the veritable drawing board. In the recent past, strategic planning called for consideration of past performance that was then projected forward based on an assumed steady-state business environment. Execution of that strategy relied upon the creation of enduring structures, the establishment of new offices, the construction of facilities, or the development of mechanical and electronic systems, which simply could not accommodate change. As a result, change had to be suppressed and avoided. Standard engineering management practice was to establish highly formalized and contractual protocols around engineering change proposals that attempted to separate change from the mainstream project activities so as to reduce their impact on current development activities.

Management practices derived from engineering's reductionist thinking and Frederick Taylor's scientific management focused on repeatability and involved relentless analysis and rationalization of work practices. The underlying assumption of this ethos was that these practices were stable and so worth the effort needed to refine them. This philosophy remains at the heart of much of today's management practices, being repackaged for new generations of managers under different labels such as quality management systems, statistical quality control, total quality management, and more recently Six Sigma, and even enterprise architecture. To ensure the scientific management paradigm was workable meant that change had to be suspended for later consideration, and considered as a minor perturbation on the principal activity being undertaken.

All that changed with the advent of software, which offered a medium so malleable [2, 13] it could be molded easily, sometimes too easily. Small changes to software code could present as vastly different system behavior and the ease with which a software developer could modify the software made for unlimited possibilities. Historically, we have never experienced such bounty, where with enough creativity and ingenuity software can create virtual worlds that connect people in new, previously unimagined ways. This innovation continues apace today, both inside and outside the enterprise, testing the fundamentals of what it is to manage.

Software development requires significantly more human interaction than traditional engineering projects. It is a more democratic process that benefits from the direct involvement of customers who can offer valuable feedback on product direction. The myriad possibilities offered by software have effectively diminished the power of the individual manager in determining the appropriate strategy to adopt. No one can be absolutely certain what will succeed in the market; it is difficult to distinguish between critical requirements, and wish lists have to be teased apart to find the compelling business features.

2.1.1.2 Agile Methods

From the outset, the application of the heavily regimented engineering management disciplines in software development was problematic [5, 6]. Software projects routinely failed to successfully deliver the required solutions within cost and schedule constraints, giving rise to the expression "software crisis."

The prevailing wisdom, represented by the familiar waterfall model, called for a contractual approach to meeting customer requirements and sequentially undertook the design, construction, and testing phases of the development. While the lock-step phases of the waterfall life cycle were never intended by Royce, who first proposed the model, the main difficulty with software lay in the very first phase where the requirements determining the scope of a project had to be agreed.

With the plethora of possibilities that software offered the customer, the burden of determining the requirements became a challenge. Defining and maintaining a stable set of requirements became a significant and ongoing challenge. In the software context, the customer, even where assisted by a business analyst, did not know or could not easily decide what they wanted. And as they developed an understanding of what a software solution could do, the nature of software meant that they *could* change their minds. Software was both a blessing and a curse; what made it so desirable also made development projects unmanageable.

The inability of engineering management approaches to deal with the fluidity of software, combined with the need to manage uncertain and changeable requirements, necessitated a new style of management. More collaborative and inclusive approaches to development were needed that recognized the critical role of the customer and the skills of a team of developers [7, 8]. At a meeting of software methodologists in Snowbird, Utah, the failure of traditional engineering practices was acknowledged in what has become known as the Agile Manifesto [9]:

Individuals and interactions over processes and tools
Working software over comprehensive documentation
Customer collaboration over contract negotiation
Responding to change over following a plan
That is, while there is value in the items on the right,
we value the items on the left more.

The challenges of software development had resulted in a management heresy. The dissatisfaction with the waterfall life cycle and the strict adherence to plan-driven engineering management practices in software development gave rise to what were termed lightweight development methodologies or agile methods. There is not a single agile method, rather a number of methods including Scrum [10], extreme programming [11], feature-driven development (FDD) [11], and dynamic systems development method (DSDM) [12] all subscribe to the principles laid down in the Agile Manifesto.

With the motto "embrace change" championed by Kent Beck [11], agile methods advocated a team-centric, managerless management approach to developing software solutions. A key way of describing agility is that it is adaptive, responding to changing conditions. Rather than lacking in discipline, agile methods often demanded more rigorous application of established best practices in order to accommodate change.

It must be made clear that few of the key principles attributed to agile methods are new and some are downright nascent. Many of the techniques to manage uncertainly had been previously defined and understood, and who could argue against greater teamwork? What agile methods, and variants such as extreme programming, achieved was in many ways the creation of a unifying marketing vehicle for a hitherto disparate group of practices. Agile methods became a brand around which the software community could rally. But it was not simply a cynical exercise to promote the separate methods that supported the agile ethos. A greater focus on key disciplines was needed if releases were to be continuous, so that the management overhead of integrating and testing each delivery could be reduced. These disciplines—such as test-driven development, but which ensured the quality of each change package, placed significant demands on software developers—were more palatable when given catchy monikers like extreme programming. Few could have imagined such enthusiastic adoption of somewhat onerous disciplines if it had been demanded of software developers by management.

2.1.1.3 Iterative Delivery

One critically important tool in the agile methods is the iterative development cycle which delivers incremental releases of the required capabilities that attempt to provide the complete solution over a number of release cycles. This approach borrowed from earlier engineering prototype development with the twist that software allowed an evolutionary approach to determining and delivering the solution without needing to discard earlier prototype releases.

Iterations effectively reduced the risk of in-flight changes and supported adaptive development by removing long delivery schedules and instead using multiple, time boxed releases to deliver a prioritized subset of the projects requirements. Object-oriented and rapid application development (RAD) methodologies advocated similar approaches for software projects with Boehm's spiral model being a popular early example of what was termed the evolutionary development life cycle.

Iterative development leverages the flexibility of software and attempts to manage requirements uncertainty through time-boxed development cycles (sprints in Scrum's idiom) and incremental releases. The aim was always to deliver working software in small, incremental units that users could interact with and provide feedback on any adjustments required. Each incremental release aims to deliver customers value while allowing product sponsors to review and direct the remainder of the development program appropriately. While no single release might make a customer happy, over a series of releases we could converge on an acceptable solution.

The word iteration has its origins in mathematical techniques that are applied where an exact solution cannot be found for an equation. Instead, iterative algorithms start with an approximate solution and repeat the process by using the result of the early iteration as a starting point. If, after successive iterations the right answer is reached, we say that the algorithm has converged on the solution. However, there is a real possibility in mathematics that a solution is NOT achieved—something that is rarely considered in agile development. Further, while iterations offer a low-risk approach to delivery, cumulatively they can be a slower, more costly means to reach the ultimate solution. Successive sprints can involve extra, albeit incremental, redesign effort which combined with the overhead of multiple releases, needs to be weighed against the benefits of iterative methods.

The direct approach to development taken by the waterfall life cycle, where it can deliver the required solution, is more efficient. By specifying the target state in the analysis phase, the development team simply works towards its delivery in a single pass of the life cycle, with no need for multiple integration phases and repeated release packaging and testing. To achieve agility and adaptability on the other hand requires software developers to work closely and utilize disciplines such as SCM with the support of automation tools. Together with the right processes and tooling, the inefficiencies of having multiple release cycles are overcome with techniques like continuous integration, which sees the automation of the software build allow continual test and adjustment of the software configuration.

Iterative development therefore compensates for its inherent inefficiency through the streamlining of the software release management process and has made a blessing of what could have been a curse. The added disciplines applied by developers allows greater visibility and confidence of what is going to be delivered, and the short release cycles combined with tool automation ensures that any changes can be readily accommodated and the software rebuilt to meet the adjusted specifications. These are the most impressive hallmarks of an efficient and effective software development team that would be well worth emulating in the delivery of strategic enterprise changes.

2.1.1.4 Challenging Management

As agile methods take root in software development, the controversy surrounding this approach has subsided even though their full implication remain to be synthesized into broader management practices. The potential deficiencies of iterative releases can be countered by demanding strongly disciplined teams and encouraging build automation, which reduces the overheads associated with a release. But perhaps most important is that these methods challenge the very role of management, distributing the traditional responsibilities of a manager. Development teams become empowered, dealing directly with the customer and prioritizing and coordinating their workload accordingly.

The rise of agile methods can be framed as part of a larger and continuing debate about the merits of management control versus that of self-managed teams and the creative freedom of practitioners. The challenge to traditional command and control management styles has a historical precedent in concurrent engineering initiatives, and given an individual's predilection for autonomy and freedom from external control, back to the dawn of humanity. The debate over management control recurs at points in history where an advance in science or technology creates new opportunities and the creation of a skilled elite. The demand for the new capability drives a power shift between those individuals who have mastered the technology and those customers or sponsors who need their skills. For example, the Cold War created a demand for the development of new weapons that outstripped the ability of the federal contractors to deliver. As a result, the government allowed innovation to flourish with the practice termed concurrent engineering [14], which can be considered a precursor to similar thinking in software that eventually found acceptance in agile methods [2].

Stages of intense innovation recur regularly, with the dot-com boom in the final years of the last century, providing a recent example. In such times, before the processes for working with the new technology are well understood and remain more an art than a technology, the pendulum swings away from management control towards autonomy and freedom for the skilled practitioner. It is at these junctures the role of management and the contribution that managers make is increasingly questioned. Generic managers unschooled in the new technologies often add little value as they do not have the understanding needed to facilitate the process. Their uncertainty and inexperience with the new technology can cause indecision and risk aversion that can act as a drag on an otherwise productive team. Further, customer demand for the new capabilities offered can make financial controls on programs unnecessary, and if the market is willing to pay a price premium then competitive advantage results for early delivery of these capabilities. At such points in time, innovation trumps control, and management can become a constraint on the exploration of new possibilities and an as-yet indeterminate future.

In the enterprise, we are facing such a point in history. The introduction of software can be seen as a disruptive technology that poses a challenge for conventional management practices and forces innovation and a reevaluation of the role of manager. The inherent flexibility of software solutions demands a more collaborative approach to development so that does not negate the need for some coordination and control. The reality that software is developed by teams rather than individuals has led naturally to the attractiveness of self-organizing or self-managing teams. While the role of the manager has been reduced to facilitation and reporting, the focus on team visibility has yielded simple and effective techniques such as the backlog and burndown charts. To achieve that, software developers had to wrest control from traditional managers in order to discover new tools that allowed them to manage their technology more appropriately.

Striking the balance between expert practitioners and the perceived overhead of management is key. While agility might be an important reminder of the need to be lean and fast, no one advocates losing all management oversight. Arguably, agile methods require greater management; it is simply that the management tasks are distributed across the team and integrated into development practices. So while the traditional role of manager is threatened, management is not diminished. However, in the enterprise context, management oversight must extend beyond a team's boundaries, to include the role of other, interdependent teams that are needed to deliver a complete enterprise solution. In a devolved environment, strategic execution is the result of the concerted effort by team of teams and so a singular focus on change is inadequate.

But while agility is a perfectly sensible approach for a new, emergent product or solution that is not well understood, it does not easily scale in organizations that have a complex preexisting, legacy environment.

2.1.2 Enterprise Complexity

While agility may have captured the zeitgeist, it has done so on the basis of some simplifying assumptions. Small, cohesive teams are proposed to improve communications and reduce the need for formal processes and documentation, while the ability to respond to change is addressed through incremental delivery and short-release cycles. Essentially, agility works best when complexity is low or at least emergent in the solution through several releases.

The distributed nature of the modern enterprise presents a number of practical problems to the creation of close-knit, colocated team environments. The trend towards greater outsourcing, offshoring, and hot desking can result in dispersed teams often working in different time zones. How efficiently each individual team operates can be less important than how their different efforts are coordinated and integrated into the desired solution.

The culture of a large organization and its associated bureaucracy can actively work against the core tenets of agility. The assumption that individuals can represent the "customers'" viewpoint is more likely to be an exception rather than the rule. The difficulty of making decisions for a collective group, whether internal users or by proxy for an end customer, can be a difficult and risky decision that many in a large organization would consider best avoided. So, instead of a clear direction on the desired solution or product vision, there can remain frustrating ambiguity and indecision. Even where committed project sponsors have the best of intentions, there is rarely the time, or the ready answers, needed to satisfy and clearly direct the project team.

As alluring as agile methods may be, the enterprise context can present real barriers to their adoption. The scale of an enterprise and the inherent organizational complexity, risk aversion, and fragmented decision-making slows decision-making and become a significant aspect of a program's schedule. Multiple review cycles, the analysis of alternative scenarios and difficulties obtaining management sign-off are all characteristics of business-IT interactions in the enterprise. The techniques that can work well at a smaller scale where individuals have the power to provide definitive guidance on requirements are not as effective in larger enterprises. The informal communications and tacit knowledge shared by a project team using storyboards in common areas cannot be relied on in the more impersonal enterprise environment. By way of example, a senior VP of a major financial company was quoted describing the limits of agile [15]:

> When you have one project on a whiteboard with Post-its, it is fine, but when you have five or six projects, the whiteboard approach just does not cut it. We haven't got enough whiteboards.

While the report went on to suggest that the answer was improved tool support, this somewhat flippant comment is indicative of the difficulties of scaling agile methods. This is not simply about the scale of individual projects, but the number of projects of varying sizes which comprise the enterprise change portfolio. It is in dealing with the multiplicity of projects, all which might indeed have adopted agile methods, that adjustment is needed to achieve some manner of agility in the enterprise. In particular, the scaling hinted at in the above quote is not of a single large project, rather it is of multiple projects peppering the enterprise with uncoordinated changes.

One of the most challenging aspects of agility in the enterprise is that few capabilities can be developed independently and more often have to co-exist with, or fit into, a wider enterprise context. Project releases have to be integrated and tested so they can merge with the as-is enterprise architecture—often first having to be merged and integrated with other concurrent projects. The autonomy of a project in the enterprise is diminished as a result, with the releases becoming the focus of delivery. These conditions will be studied in Chapter 3, and the importance of agility at the project level is shown to be less relevant than it is at the portfolio level.

The premise at the heart of ERM is that as the modern enterprise is a continually adapting, evolving complex software system, then the management experiences and techniques in managing software should cross over to manage enterprise evolution. But in order to provide workable answers for managing enterprise evolution, there is a need to first understand the management implications of growing complexity, the resistance it places on change, and the consequences of shared architectures.

2.1.2.1 Enabling Enterprise Technologies

Looking at how the term enterprise is used by products and in the media, one can be forgiven for thinking that the enterprise itself is a manifestation of the IT field. Marketing literature for IT products routinely appropriate the word to describe their most powerful and advanced offerings. Development and IT management best practices use the term enterprise to describe the deployment of a discipline or toolset consistently across an organization. So we find the term enterprise project management, enterprise content management, and enterprise risk management including earlier manifestations such as enterprise resource planning (ERP), today simply termed enterprise systems, to reflect their breadth of coverage. And now this book is introducing enterprise release management.

There is a case to be made that in many ways the modern enterprise is a creation of the widespread availability of IT solutions that have enabled the integration and coalescing of business functions through support for improved communications, process automation, and information sharing. Arguably, the implementation of new, software-based, digital technologies provides the modern enterprise the capabilities to grow and be more efficient than at any time in history. IT has given rise to several generations of enterprise systems which, from the earliest incarnation of ERP systems, have created the vision of integrated and automated business operations.

The possibilities presented by IT capabilities captured the imagination of business executives from the outset, even as years on, the IT industry struggles to deliver on their promises. Yet, for all the challenges of making IT meet strategic objectives, IT has been the spur for business innovation and a growing part of enterprise operational expenditure. However, IT alone is not the solution as evident when considering the dimensions of enterprise architecture. So by extension, neither can it be entirely the problem.

Increasingly sophisticated enterprise systems offered by vendors like Oracle, SAP, and Microsoft that continue to grow in scope and capabilities, providing a digital platform [4] for businesses, and with the people and processes that bring them to life, have enabled the integration of the separate functions and more streamlined operational processes across multiple lines of business. The enterprise is now connected both within and without, lessening the need for middle management and reducing, if not eliminating, functional silos with capabilities including straight-through-processing (STP) that more directly connect customers to back-office systems. Leading businesses have applied Web and Internet technologies to tame and orchestrate complex supply chains that dissolve old boundaries and merge previously distinct functions into an integrated process.

Implementation of such technologies have, literally, enabled the enterprise by integrating what might otherwise be a collection of separate, less-efficient business units, each carrying overheads which can be performed by core systems and services in the enterprise. But, at the same time, an integrated enterprise encounters new challenges in managing a shared enterprise architecture. The investments made by IT departments in enterprise systems, whether intended to grow business, expand market reach, or streamline business processes, have now climbed to consume 14% of operational expenditure in the government sector and 13% of the financial services industry [6]. The rapid rate at which new technology innovations move through the hype cycle [16] and go from inflated expectations to becoming a legacy system has been a hallmark of the digital era. In the current phase of being digital, enterprises have to contend with the continual change, whether refinement or replacement of IT solutions, while business operations become reliant on an interconnected configuration of systems that only grows more complex as an enterprise evolves and matures.

Cynthia Rettig [17] caused controversy when she argued that enterprise systems had become too complex and could indeed present a risk to business. Indeed, the article challenges the effectiveness of the current best hope for enterprise simplification: service-oriented architectures (SOA). In somewhat colorful language, Rettig voiced the frustration felt by many IT practitioners and business executives.

> The triumphant vision many buy into is that enterprise software in a large organization is fully integrated and intelligently controls infinitely complex business processes while remaining flexible enough to adapt to changing business needs ... While a few companies like Wal-Mart Stores Inc. have achieved something close to that ideal, the way most large organizations actually process information belies that glorious vision and reveals a looking-glass world, where everything is in fact the opposite of what one might expect. Back-office systems—including both software applications and the data they process—are a variegated patchwork of systems, containing 50 or more databases and hundreds of separate software programs installed over decades and interconnected by idiosyncratic, Byzantine and poorly documented complexity.

While the enterprise systems that power most major corporations are clearly imperfect, they are increasingly ubiquitous. Businesses cannot easily decommission older piecemeal attempts by IT to automate manual business processes and it is common for stakeholders to demand redundant or superseded systems to be maintained, even as new, more integrated IT offerings are put in place. Few enterprises have the opportunity to completely start afresh and transform their businesses with a wholly new enterprise system. Instead the gradual, somewhat glacial transitions from older technologies to new, integrated solutions are commonly the daily fare for enterprise architects. To figuratively lift the lid on most enterprises' IT architecture is likely to reveal not the efficient, clockwork mechanism executives hoped for, but rather a dystopia of loosely interconnected processes and legacy systems connected together with the software equivalent of rubber bands. The fact is that what software developers term "spaghetti code" or "big ball of mud" architectures have now achieved enterprise proportions.

2.1.2.2 Growth, Complexity, and Friction

There is inevitability complexity in the enterprise. Simply being in business over a period of time sees an enterprise's architecture grow through accretion. Like any successful product, an enterprise, even where it starts out small, independent, and agile, has an inexorable tendency to increasing complexity.

While the business may be oblivious to the management burden associated with maintaining disparate systems, the strategic focus cannot simply be on new capabilities. As unproductive as it seems from a business perspective, resources have to be assigned and management attention needs to be paid to essential, less visible, maintenance activities. Complexity drives up costs, from the management and support personnel needed to the maintenance of the facilities and IT infrastructure that contribute to the growth in a business' operational expenditure.

This is the predicament that IT departments can suffer in the enterprise: having great technical complexity yet being seen by the business simply as a utility which, while indispensable, provides little business differentiation. Nicholas Carr asked the insightful question in his book: "Does IT Matter?" [18]. The very success of IT in becoming an integral and enabling part of the business brings with it the seeds of its demise. With commercial enterprise systems becoming ubiquitous, businesses question the contribution and strategic advantage offered by IT departments—leading to cost reduction efforts that can result in outsourcing and/or offshoring to the lowest bidder.

The fact that any operating business or enterprise has an as-is architecture at all is a problem for enterprise change. Best practices such as ITIL and COBIT [19] focus on the management of smaller IT changes to the operational environment while non-IT changes and larger-scale, strategic changes are kept and managed separately. The separation of strategic from operational change has allowed idealistic development and delivery models to persist in the charade that most projects are greenfield efforts that start with a blank canvas. Recognizing the reality of an as-is architecture from the outset both reinforces the fact that most projects in the enterprise are vehicles for delivering change, and raise a warning flag as to what will and won't work in such environments.

There is little question that agile methods work best in greenfield projects seeking to explore and discover the possibilities that cleverly crafted software solutions can offer business. To be agile is to be small, focused, and fast, moving as required and free of the encumbrances of the enterprise. Without a firm "plan" to direct them, an agile team can set the product course and define the to-be architecture as it sees appropriate. As an independent initiative, there is no need to comply with organizational standards, security policies, and the like that can distract the team from the exploring the means to offer solutions that maximize business value.

Expedient decisions taken earlier to deliver new functionality quickly may need to be reappraised in light of new requirements and a growing usage of the system. As a result, refactoring is an agile technique that ensures design rework becomes an integral part of the development process essential to rearchitect the solution to address growing complexity. This effort can naturally slow down the development of new features but it is the trade-off that Agile Methods make and essentially prioritizes change over complexity. Many would argue that the need for refactoring, to scale a solution due to increasing usage and patronage, are the kind of problems that business customers want! Better to have the priority on useful features with a simple architecture, than to have a scalable, robust architecture with no patronage.

The point is not to debate the pros and cons of different methodological perspectives but to recognize that at the very heart of the need, and success, of agile methods is the fact that in elevating and embracing change, complexity has been deprioritized and considered a secondary consideration. Yet, what we can expect from this is that the benefits of a freewheeling agile development will in of itself eventually begin to slow as the solution evolves and refactoring efforts being to grind down progress.

Complexity acts as a frictional force against change that eventually slows innovation. Even in the context of a single project, agile methods can make great strides in delivering early solution capabilities, but the development velocity; that is, the number of new capabilities delivered per unit of time, inevitably slows.

The challenges of scaling agile are compounded in the enterprise context. IT organizations use standardized operating environments in order to reduce support and infrastructure costs that can be seen to stymie agility. What starts as a simple and independent solution can quickly the need to has to adopt more onerous enterprise standards before it can be integrated into the as-is architecture and a part of the operational environment that has the support of the IT organization. The need to meet IT security standards for example can introduce new, cross-cutting requirements that can take a simple application into a new architectural league. At some point, an agile project in the enterprise will find that the next iteration of the solution it offers can no longer be determined independent of external enterprise considerations. It is at this point when agility and innovative changes hit the constraints of enterprise complexity that the effectiveness of agile methods wanes.

2.2 Managing Change and Complexity

As an enterprise grows, the pursuit of economies of scale drives the sharing of common processes and technologies increasing interdependencies and complexity. This serves to make the enterprise less agile then we would like, a reality that is not always popular with business executives who see only opportunities in change and discount the implication of a complex enterprise architecture.

While agile methods, in their current form represent the ideal case of "frictionless" development, in reality, complexity will always be present, or emerge, to dampen the benefits of agility. A more robust theory of program delivery is needed; one that addresses the dynamics of change and complexity at a level of abstraction that scales to the enterprise.

2.2.1 Change-Complexity Cycle

To successfully deliver strategy in the modern, federated enterprise requires that management recognize change and complexity as closely coupled forces that cannot be addressed in isolation. Responsible management does not attempt to dismiss the unpleasant reality of enterprise complexity but rather embrace it to ensure a reasonable business outcome is achieved.

Much of the debate between business and IT can be viewed as the parties arguing for one or the other side of the change and complexity question. Business executives seeking a swift response to competitive opportunities prize agility and demand support services such as IT deliver on short-term requirements at the expense of future efficiencies. This is the so-called business imperative that requires IT to be aligned with the business and prizes agility and responsiveness. In contrast, an efficient IT organization operates using traditional engineering-based disciplines that focus on managing and maintaining the integrity of (in this instance) the enterprise architecture. Their aim is to ensure that the systems perform reliably, IT services are provided within acceptable service levels and operations are cost-effective.

The above discussion represents the different perspectives of business and IT, the consequences of which will be discussed in Chapter 6. Neither of these views are necessarily correct, with individuals essentially playing roles they are assigned. Technologists can equally be innovators who challenge the boundaries of existing capabilities, while business executives with financial responsibilities would seek efficiencies that come from standardized architectures. While each viewpoint is relevant, different business conditions call for one or the other of the above positions to be prioritized over the other. In the longer term there are consequences arising from each decision and so in a holistic context the relationships between change and complexity can be seen as cyclic.

Figure 2.2 illustrates the implications of an enterprise facing an excess of either change or complexity. When the organization is unable to accommodate or balance these twin forces, the result can be the generation of the opposite condition. In the context of enterprise strategy, changes can be considered those

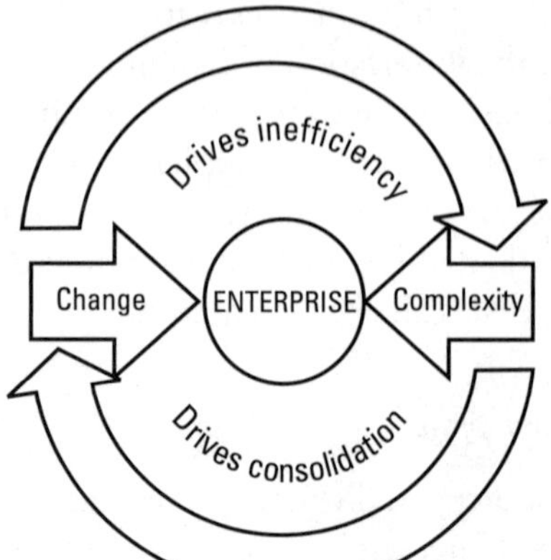

Figure 2.2 Change and complexity cycle.

strategic initiatives sponsored by the business customers, while technology's role is to manage complexity and ensure IT products and services are supplied cost effectively. While this customer-supplier (or is it master-servant) perspective is not palatable to many IT practitioners, this description is a convenience to illustrate the more fundamental relationship between change and complexity.

2.2.1.1 Innovation and Alignment

In an attempt to meet competitive demands and provide new products and services that customers want, business executives routinely prioritize in favor of innovation and change, irrespective of the downstream consequences. In the words of management guru Peter Drucker [20]:

> A good manager always focuses his attention more strongly on opportunities than on risks ... constantly worrying about problems does not really move things forward.

Poorly applied, the result of such thinking is that business-IT alignment requires IT to marshal its resources so it can support and quickly respond to changes in the business strategy. While this is fine in principle, problems arise when IT and the other business support functions are not able to cope with the demand for changes to the enterprise architecture.

The focus on alignment with business imperatives has led to IT managers making compromises that meet business priorities at the expense of architectural design inefficiencies that give rise to greater complexity. As a shared resource across the enterprise, IT has borne the brunt of this more general enterprise management dilemma. The alinement trap [21] will be explored further in Chapter 6, but its cause can be summarized as follows:

> IT's effort to satisfy its various (and sometime conflicting) business constituencies created a set of Byzantine, over-lapping systems that might satisfy individual units for a while but did not advance the company's business as a whole.

Business-IT alignment can be seen to prioritize in favor of business agility and time-to-market while sacrificing, or at least postponing, operational business and IT efficiency. An overemphasis on change does not allow sufficient time for the enterprise to adapt and develop the appropriate processes and governance structures needed, resulting in wasteful disorganization, duplication, and redundancy. This in turn can result in the need for future, corrective changes and refinements that can create continual churn that stresses operational resources and unsettles the organization.

In the drive to transform the modern business and adjust to a changing business environment, compromises have to be made that can serve to increase overall operational complexity. An enterprise's architecture can degrade by having new processes or systems not well integrated into core platforms, further adding to support and management costs.

2.2.1.2 Simplification and Consolidation

The alternate path in the cyclic relationship between change and complexity is where the latter becomes unsustainable and a simplified enterprise architecture becomes imperative.

This path in the cycle represents in many ways management's capitulation in the face of growing enterprise complexity. Solutions offered to address the complexity can take many forms, but the options are often based on the notion that the management challenge can be passed on to an expert third party. Currently favored remedies include outsourcing or offshoring, usually in some combination that provides the benefits both of separation of management responsibility and desirable cost reductions.

With IT often being the first to exhibit symptoms of growing complexity, the latest solution proffered by IT vendors are regularly positioned as the best means of achieving standardized processes and efficiencies. Simplification and consolidation is promised by the vendors of enterprise systems that offer to transform the organization by automating best-practice business processes. They offer the hope of escaping the redundancy and waste that can accompany uncontrolled growth with a fresh, clean-slate approach geared to the meet the enterprise's future needs.

Anyone who recognizes that the enterprise architecture goes beyond simply the technology dimensions might have reservations over the brave new order that the above solutions offer. Passing the "complexity buck" to an entity outside the enterprise, whether service provider or vendor, does little to directly reduce or control complexity. In fact, introducing new IT systems and suppliers into the enterprise mix can serve to exacerbate the problems. Instead of the promised benefits, the stark reality is that there is no simple remedy to the complexity problems—we simply have to embrace it and, as suggested by a recent CIO survey [22], attempt to capitalize on complexity by managing it better than our competitiors.:

While resolving problems with complexity have traditionally used divide and conquer techniques, the more recent embrace of change now presents us with a new dimension to the problem. Decomposition alone will simply yield an integration nightmare with versions of business and technology components that can confuse strategic initiatives.

The relationship between change and complexity is neither a virtuous nor a vicious cycle—it simply is a consequence of an enterprise having a limited ability to effectively absorb change, and on the other hand, cope with and manage complexity. The cycle cannot be effectively managed by treating each dimension in isolation. New paradigms and techniques are needed that transcend the cycle rather than simply trading off one for the other. A better understanding the dynamics of the change-complexity cycle may at least be able to attenuate its effects by addressing root causes.

2.2.2.2 Mapping the Management Response

In the relatively brief history of formalized management thinking it is not surprising that change and complexity have been recurring, albeit not always recognized, themes. Dependent upon the business conditions, the introduction of new, disruptive technologies can be the catalyst for the rejection of management control in favor of a more free-flowing innovation cycle.

It is informative to assess different management approaches in terms of their ability to address complexity and change in the enterprise. In order to do that, we attempt to categorize or map the traditional, plan-driven disciplines and the modern agile approaches to illustrate their management "sweet spot" and identify what remains to be addressed.

To facilitate that, we employ the classic four-square grid and position the opposing and orthogonal forces of change and complexity on separate axes. The quadrants provide a simple means for categorizing different business conditions by the level of change and complexity that is present in the enterprise and the four combination of low-low, low-high, high-low, and high-high can be visually displayed.

Figure 2.3 illustrates the coverage of the different management approaches based on earlier discussion of their expressed focus. The fact that agile methods are best suited to managing the change dimension shows it spanning quadrants 1 and 2. The previously discussed difficulty of scaling the agile philosophy to accommodate increasing enterprise complexity has meant that it is less suitable for environments found in quadrants 3 and 4. Correspondingly, the plan-driven, top-down management approach addresses the complexity dimension and so spans quadrants 1 and 3. These cover environments that require a defined architecture to enable the decomposition of large programs of work that can then be allocated to multiple teams working to strict, contractual interfaces. As such this engineering-based approach is less tolerant of high levels of change, making it less suited in quadrants 2 and 4.

The first quadrant representing the simplest business environment, characterized by low change and low complexity, is the only quadrant where the two management paradigms overlap. In such a simple enterprise we can frankly expect any management approaches to work. The application of basic management principles should yield success given the lack of any major challenge.

Much of the debate regarding the merits or otherwise of the two management approaches discussed are meaningless without reference to the nature of business environment. What the quadrants in this model illustrate is that

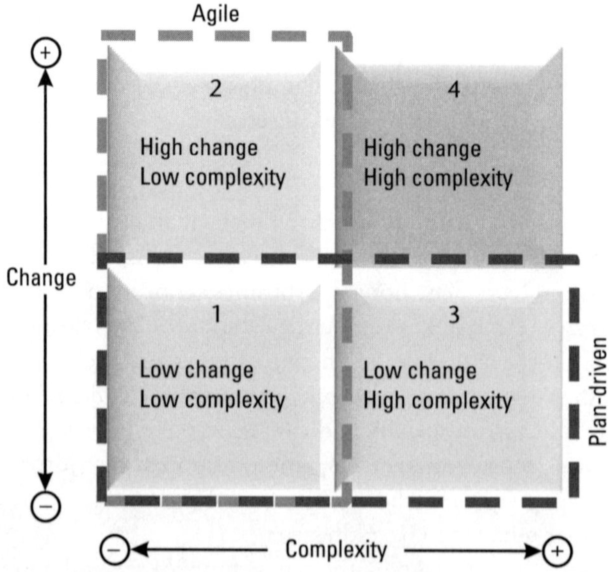

Figure 2.3 Sweet spots for agile and plan-driven approaches.

each approach has its comfort zone where it can be expected to be successful. The only real choice that is available is for those simple projects, where both approaches could be successful and the decision does not warrant too much debate.

The important fact that Figure 2.3 vividly illustrates, one that is clear when the argument is couched in terms of change and complexity, is that the two management approaches that are our current operating paradigms are insufficient. There is a gap in our current management arsenal when it comes to managing environments characterized by both change and complexity—shown in the figure as the fourth quadrant.

Current management practices focus on, and so can only cope with, one of the two dimensions of change and complexity at a time. When confronted with both of these forces, management has attempted to constrain one while managing only the other. Traditional, plan-driven management attempts to deny change and manage it separately through a formalized change control process that can slow its response to change. Agile management on the other hand repudiates complexity and instead feigns simplicity through iterations and systematized rework it redefines as refactoring. Essentially, when either of these approaches operates outside its sweet spot, it encounters difficulties that have to be accommodated somewhat artificially.

Unfortunately, most modern enterprises find themselves operating in the fourth quadrant—precisely where our management approaches are weakest. As such, this quadrant represents an open challenge for the delivery of strategic outcomes in the enterprise. It requires new, more integrated management tools and techniques that embrace both change and complexity. And it is the interplay between these two fundamental forces operating in the enterprise that frames the arguments that lead to ERM.

References

[1] Friedman, T. L., *The World is Flat 3.0: A Brief History of the Twenty-First Century,* New York: Picador, 2007.

[2] Blackburn, J. D., G. Hoedemaker,and L. N. van Wassenhove, "Concurrent Software Engineering: Prospects and Pitfalls," *IEEE Transactions on Engineering Management,* Vol. 43, 1996, pp. 179–188.

[3] Campbell-Kelly, M., *From Airline Reservations to Sonic the Hedgehog: A History of the Software Industry* . Cambridge, MA: The MIT Press, 2004.

[4] Weill, P., and J. W. Ross, *IT Savvy: What Top Executives Must Know to Go from Pain to Gain,* Boston, MA: Harvard Business Press, 2009.

[5] Brooks, F. P., *The Mythical Man Month,* Reading, MA: Addison-Wesley, 1995.

[6] Gartner Research, "Information Technology Key Metrics Data," Gartner Inc., 2009. http://www.gartner.com/.

[7] Highsmith, J. A., *Adaptive Software Development: A Collaborative Approach to Managing Complex Systems.,* New York: Dorset House, 1999.

[8] Cockburn, A., *Agile Software Development: The Cooperative Game,* 2nd ed., Reading, MA: Addison-Wesley Professional, 2006.

[9] Agile Alliance, "The Agile Manifesto," 2001, http://www.agilemanifesto.org/.

[10] Schwaber, K., *Agile Project Management with Scrum,* Seattle, WA: Microsoft Press, 2004.

[11] Beck, K., *Extreme Programming Explained: Embrace Change,* Reading, MA: Addison-Wesley Professional, 1999.

[12] Coad, P., E. Lefbvre, and J. De Luca, *Java Modeling in Color with UML: Enterprise Components and Process,* Upper Saddle River, NJ: Prentice Hall International, 1999.

[13] Stapleton, J., and DSDM Consortium, *DSDM: Business Focused Development,* 2nd ed., Boston, MA: Pearson Education, 2003.

[14] Wheeler, R., R. W. Burnett, and A. Rosenblatt, "Concurrent Engineering: Success Stories in Instrumentation, Communications," *IEEE Spectrum,* Vol. 28, 1991, pp. 32–37.

[15] *Forester Wave, Agile Development Management Tools,* Q2, 2010

[16] Hype Cycle, Gartner Research Methodologies, Gartner Inc., 2011 (cited June 17, 2011), http://www.gartner.com/technology/research/methodologies/hype-cycle.jsp.

[17] Rettig, C., "The Trouble with Enterprise Software," *MIT Sloan Management Review,* Vol. 49, Fall 2007.

[18] Carr, N. G., *Does IT Matter? Information Technology and the Corrosion of Competitive Advantage,* Boston, MA: Harvard Business School Press, 2004.

[19] COBIT 4.1, *Control Objectives for Information and Related Technologies,* IT Governance Institute, 2008.

[20] Drucker, P. F., *The Effective Executive*, Harper Collins Publishers, 1993

[21] Shpilberg, D., et al., "Avoiding the Alignment Trap in IT," *MIT Sloan Management Review,* Vol. 49, Fall 2007.

[22] Institure of Business Value, *Capitalizing on Complexity*, IBM Corpration, 2010.

3

The Enterprise Pattern

Everything should be made as simple as possible, but not simpler.
—Albert Einstein

The challenge of executing strategic change, while evident to many practitioners, can benefit from further clarification. This chapter explores the characteristics that define enterprise evolution in the fourth quadrant of Figure 2.3. The combination of change and complexity that is present in such an environment is shown to give rise to what we describe as the enterprise pattern.

The pattern identifies the interdependencies that arise when a change portfolio comprising of several concurrent change initiatives impacts a complex architecture. The scenario, while familiar, is shown to be ill-suited to both agile methods and their adaptive, self-organizing principles, as well as traditional management and its emphasis on planning and control. Instead the management challenges inherent in this pattern can be attributed to the enterprise exhibiting the behaviours more commonly found in complex adaptive systems.

Representing the relationship between change and complexity as a pattern is appropriate as it will be shown to be evident across several business domains. Like the Russian dolls portrayed in Figure 3.1, the pattern can scale from small teams to value chains that encompass partner organizations. While the pattern is to be found within complex businesses undergoing transformation, it is best understood and closest to being mastered in tight-knit software development teams working with complex product architectures.

The enterprise pattern will be shown to be closely related to the management challenge found in the fourth quadrant, describing the problem space that is poorly addressed by existing management approaches. The pattern provides a useful diagnostic tool that helps make explicit these symptoms and the

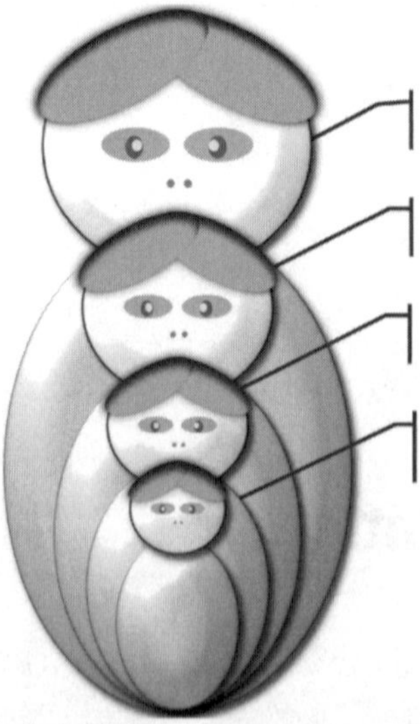

Figure 3.1 Enterprise scope and scale.

real-world scenarios that arise when tackling the somewhat abstract concepts change, and complexity. Those readers that can recognize the pattern in their own organization or from past experience will find the notation and techniques which we collectively term ERM most relevant.

3.1 Managing Enterprise Dynamics

The earlier chapters made the case for treating the enterprise as a continually evolving and adapting entity. In such an environment, implementing integrated IT solutions that aim to automate key aspects of business is the easy part. A relentless stream of changes have to be accommodated, which, if not adequately addressed across all dimensions of the enterprise, can soon erode the benefits of even the most advanced enterprise systems.

Envisioning future releases of the enterprise recognizes its architecture as transitioning in the face of inevitable changes. Even while technologists look to adopt more flexible architectures like SOA as a means of insulating the business and interfacing systems from internal changes, they are limited to the technology domain. When the nature and impacts of changes are viewed from a holistic perspective, it is the enterprise architecture—strategies, people, processes, and infrastructure that have to be considered along with technology. As in the Einstein quote that introduces this chapter, we should not be fooled into thinking that management can be effective if it is in fact unrealistically simplistic.

The somewhat naive illustration in Figure 3.2 presents the interaction of change and enterprise complexity and its simplicity belies the significance of the management challenge it represents. But in many ways this is no new problem, rather it is a different presentation and rescaling of a familiar pattern found, and solved, in systems and software engineering. It is in fact an echo of the change and configuration management (CCM) challenge reprised at the enterprise level with the nuance necessary to manage an organizational system instead of a hard-engineering configuration. This generalization sees change management as a generic, with CCM and organizational change management (OCM) as variants, each dealing with a specific dimension of the enterprise—presenting the possibility of each informing the other.

ERM then is about adapting solutions that have been shown to work in the governance of changes to large-scale engineering programs, adjusted appropriately for the enterprise scale and human interactions that dominate management activities in the enterprise. As the enterprise architecture encompasses strategy and organizational structure, somewhat fickle human behaviors and interactions play a critical role in the dynamics of an enterprise. Yet for all these distinctions, the enterprise can be seen to operate as a system and follow similar abstract patterns and abide by common principles.

Taking the systems perspective has been a tradition in management with thought leaders including Winslow Taylor, W. E. Deming, Peter Drucker, Russell Ackoff, Peter Senge, and Michael Hammer, all having started their careers in engineer related fields. Therefore, it should not be surprising that the lessons of systems engineering and software development can be repurposed and applied in the larger, enterprise context. With the ubiquity of software-intensive IT solutions today essentially running much of the enterprise, adoption of a more agile and incremental approach to determining and executing strategy seems suited to the times.

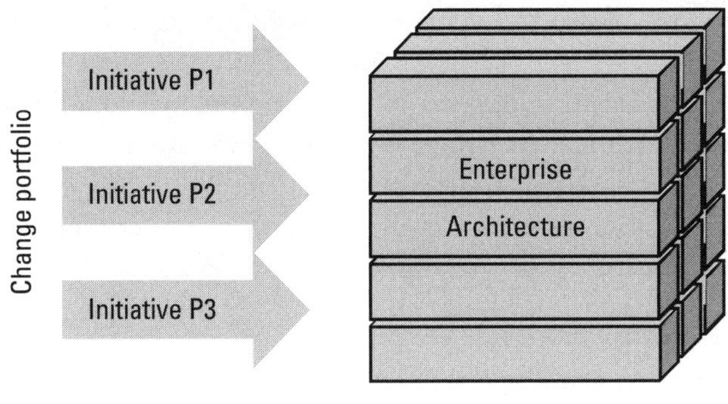

Figure 3.2 Change portfolio impacting enterprise architecture.

3.1.1 The Real Enterprise System

The kernel of ERM's agile approach to delivering strategic change lies in the recognition of the enterprise as a complex adaptive system built to achieve strategic business goals. This realization came to Ackoff and Emery [2], early proponents of systems thinking, even before the capability of IT systems to support and automate business processes was fully recognized. The "real" enterprise system should of course comprise all dimensions of the enterprise architecture, not simply the IT systems that have audaciously appropriated the term "enterprise system."

The essence of a system is that its behavior is more than simply that of its constituent parts. The performance of the whole is more a result of sometimes subtle and unpredictable interactions between these parts. While the traditional management mindset favors reductionist thinking, attempting to apply the divide and conquer principle to handle enterprise complexity—the reality can be a little more complicated. Modularity and componentization are a natural answer to complexity; however, in solution delivery it can simply move the problem downstream, creating instead an integration challenge. Systems thinkers such as Senge have argued for greater sophistication in management and challenge us to think holistically and identify causal patterns instead of simply addressing specific problems:

> *Systems thinking is a discipline for seeing wholes. It is a framework for seeing interrelationships rather than things, for seeing patterns of change rather than static snapshots.*
> —Peter Senge

While reductionism might work for a manager's understanding of a problem, it does not necessarily help in the execution of the solution. The simple hierarchies found in organizational structures are a management convenience that simply cannot mirror the more complex interactions that occur in reality. An organizational structure is in fact a model of reality, a simple and convenient static projection of a more dynamic reality—which goes some way to explaining why so many organizations routinely restructure in a vain attempt at finding the "right" structure. Sadly, there is no perfect structure, at least not one that is as simple a construct as the management hierarchy. Considered as a system, the enterprise needs a more sophisticated model to understand both the why and the how of managing the dynamic enterprise system.

3.1.1.1 The Enterprise Purpose

As an artificial, man-made construct, the enterprise requires teams of individuals, and often teams of teams, to act in a concerted and coordinated fashion to deliver the enterprise goals. The thing that aids in pulling together the disparate interests in any organization is what Ackoff termed ideals, which made for a purposeful system that required collective, human effort outcomes.

The key point relevant to the determination and execution of strategy is that such a system comprises individuals who intentionally and collectively formulate objectives, and yet are ideal-seeking; choosing alternate objectives that more closely approximates its idea. It is this high-minded, collective action that makes enterprise strategy a noble endeavor. Indeed, Ackoff and Emery note that, "The capability of seeking ideals may well be a characteristic that distinguishes man from anything he can make, including computers" [2].

The "real" enterprise system can therefore be viewed as a collection of teams and groups of individuals sharing an ideal and working together to achieve common goals. For many traditional businesses, that goal is to offer shareholder value, although some enterprises, such as not-for-profit organizations, can aim for higher ideals beyond simply financial considerations. Whatever the ultimate purpose, organizations are designed to deliver the agreed goals and bring together the different expertise and specialist functions necessary to achieve that end.

In essence, the mission, or strategy, of the enterprise needs to ensure that the different stakeholders and various lines of business across the organization are aligned with a shared and uniting purpose. Pragmatically, differing short-term objectives will result in negotiation and compromise between the different parties as this strategy is executed. So strategic determination and execution will necessarily require governance and oversight to overcome shot-term thinking and the resulting dissonance.

That is particularly the case when strategy increasingly is about changing and transforming the business, rather than expanding or improving performance in the existing businesses. Fewer conflict scenarios arise when strategy aims to seek new markets, ramp up production, or grow the sales force; such initiatives do not fundamentally change the operating model of the business. The enterprise system essentially remains stable even if it is tweaked, ramped-up, or replicated—its architecture remains essentially familiar and recognizable.

In earlier, halcyon times businesses could rely on this stability to get a return on their investments in their business model over the long term. That model, the enterprise system, could be defined in detail without the fear of frequent changes causing updates to become onerous and costly. The assumption of a steady-state and the expectation of stability over the life of the business meant that architectures could be described comprehensively. It made sense to invest in quality management systems that defined control processes in detail and initiate performance improvements programs to hone the business model.

Figure 3.3 shows a schematic of a traditional "static" architecture, named to distinguish it from the transformational aspects of change that will be discussed shortly. The component inventory (which can be viewed as a system inventory or portfolio when considered from an IT perspective) comprises

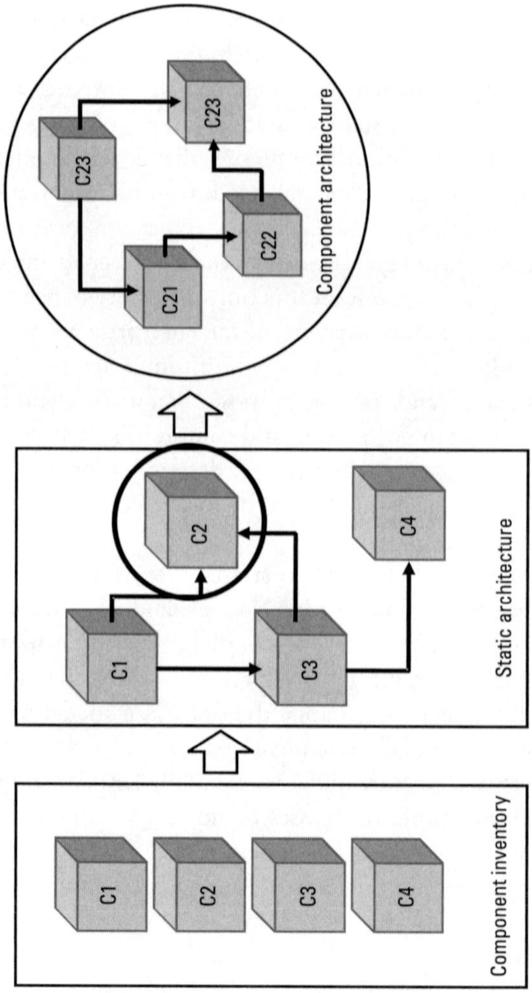

Figure 3.3 Traditional "static" decomposition of architecture.

elements of the operational architecture (denoted as C1 to C4) that interact in some way to support the business model. Each component, irrespective of whether it relates to the organizational, business, or technology dimension, can be further decomposed into a more detailed representation termed the component architecture, as shown in the figure for the C2 component. While the desirability of defining this static architecture is undeniable, the benefits can be debatable where the stability assumption breaks down. The overheads of maintaining comprehensive architectural descriptions can become significant. Like the plans and documentation that agile methods challenged, architecture becomes a potentially wasteful exercise when faced with continual change.

As steady-state conditions become less common in modern enterprises where competitive strategies demand innovation, keeping the business operating model continually in flux. In this new order, the enterprise architecture too is continually adapting, as best it can, to new market conditions. As the enterprise evolves, new business units can start up, others merge, and some simply fade away into low-margin legacy operations. During such transformations leaders and executives must reset the enterprise mission and make clear the strategy that contextualizes these changes. It is this more dynamic picture of the enterprise that can benefit from the ERM with tools and techniques better suited to visualizing the evolving enterprise. The focus on agility comes at the enterprise level by defining a coherent set of transition architectures that correspond to enterprise releases, rather than investing too heavily in describing static future states.

3.1.1.2 The Dynamic Enterprise

Describing an enterprise as a complex adaptive system can be an invitation to consider highly sophisticated models that have been developed by scholars and researchers attempting to understand their behavior across a range of physical, biological, and technological domains. Yet a management perspective can, and has, in works like Senge's, found value in simpler abstractions that nonetheless captured an essential reality not previously considered or accepted. To be useful in practice, managers need to embrace just enough of a complex reality in order to attain an insight into the problem space—things only need be as complex as they need to be.

In fact, the implications of the dynamic enterprise should be clear to everyone: continual change has made it difficult to maintain a systems or architectural view of the enterprise. As such, many disciplines and best practices find it difficult to engage and perform their duties given the pace of change in the organization. However, the insight that needs to be tackled by the next generation of management thinking, that must go beyond the apparent contradictions of agility and adaptation versus planning and control, is that changes cannot be considered in isolation of their systematic impacts.

Focusing on the temporal dependencies that become apparent in the dynamic enterprise make it easier to understand what it is that needs to be managed, without getting distracted by the complexity of the static architecture. Focusing directly on the changes shows that traditional attempts at defining and deconstructing architecture is not necessary to understand change execution—which is of course why it can so easily be neglected. However, a different problem emerges, one that is not usually given the designation of a discipline but needless has to be performed. This is the domain of release management, which is evident when the impact or implication of changes on the whole system (or product) is considered holistically.

Emergent behavior, where new properties arise when components are brought together, is one of the central tenets of complex systems. Numerous examples of the such behavior can be found in nature and society, and this thinking underpins key aspects of agile thinking and its belief in self-organizing principles. From a systems management viewpoint however, it should be clear that attempting to understand emergent behavior without addressing the system as a whole is bound to be ineffectual. How can a problem that occurs in the large be addressed in the small where it is not even evident?

Figure 3.4 represents what we term the enterprise pattern and details the notion illustrated in Figure 3.2, where changes are designated as separate

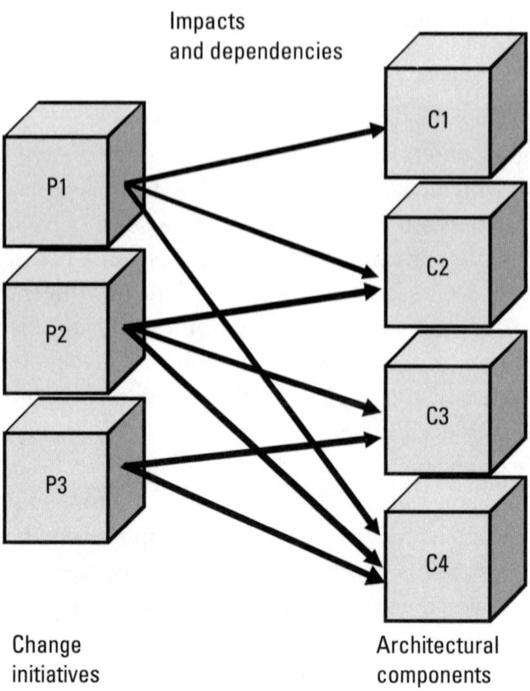

Figure 3.4 Example of the enterprise pattern.

projects or programs impacting various, unspecified components of the enterprise architecture. The pattern shows the potential many-to-many impact relationships across the three change initiatives identified as P1, P2, and P3, impacting components identified as C1, C2, C3, and C4. These could represent IT systems or business functions that have to either be created, modified, or enhanced depending on the requirements of the changes.

While the figure is simple almost to the point of simplistic, it will be shown to exhibit sufficient system properties to illustrate the challenges associated with managing both change and complexity. The initiatives on the left may appear to be isolated and separate changes to their sponsor and stakeholders, but the fact that they impose requirements for related capabilities and impact core business systems and processes means that they form now a tightly coupled portfolio—one that will be shown to present new challenges for management.

While the scale of the change initiatives and the architectural components can and will vary dependent upon the nature and scale of the organization, the pattern has intriguing implications for the management and delivery of enterprise change. Considering the multiple impacts that these change can have on a complex, shared enterprise, architecture serves to join the dots and directly connect the disciplines of program and project management (PPM) with EA. This link is provided by the release management, which is often omitted at the enterprise level. Instead, distinct and disconnected practices separately address strategic planning and execution, all working to deliver business value but working to different time-horizons.

Yet these conditions are familiar and the problem of multiple teams working across the delivery life cycle is well understood, if not fully resolved, in the software domain. At a lower level of granularity the enterprise pattern is also evident in that domain, with software releases being an example of the bridge between what is planned for the future and what has to be delivered today. To better manage strategic execution we can leverage experiences and lessons learned in software development and apply them in the enterprise.

3.1.2 Managing Change Dynamics

The many-to-many pattern shown in Figure 3.4 occurs routinely in software development, where user requests often require multiple changes to source code. The correspondence to the enterprise situation is even closer where the pattern is evident at higher abstraction levels when component-based development (CBD) or SOA is adopted.

Clearly there is a continuum that scales from the simple changes to source files made by an individual developer, to the enterprise pattern that may involve multiple transformation programs requiring modifications to several business functions and their supporting architectural components. This symmetry exists because complex software, like the enterprise, exhibits system behaviors and, as previously discussed, is in fact an enabler of greater business flexibility. The implication is that we can benefit from the insights gained in software and establish a more unified approach to managing changes at varying levels of granularity and abstraction.

3.1.2.1 Managing Software Change

In software, addressing change *and* architectural complexity is critical to delivering a quality solution as the software medium is so amorphous that it virtually attracts change. This makes maintaining the integrity of a complex software configuration as it evolves a recurring theme in software management.

While modifications to a software system are easy to affect, there is the potential for apparently minor and inconsequential changes to have a disproportionate impact that can result in catastrophic disruption or failure. All that differentiates working software from a useless collection of source files is its *configuration,* which is the result of careful change and configuration management. As a result, the discipline of software configuration management (SCM) has achieved a maturity that was not even possible in the systems engineering domain that first gave rise to CCM, which predates it by decades.

The challenges associated with the enterprise pattern have been tackled over the years in software development and yielded in a rich set of principles, techniques, and tools. SCM tools in particular have recognized the value of capturing the change to provide a context for the actual execution of that change. This realization resonates up the layers of abstraction and can be viewed as programs and projects providing the context for business and IT changes, to the ultimate abstraction where strategy provides the context for enterprise evolution.

Rational Software, a division of IBM, and a leading vendor of SCM tools, coined the expression Unified Change Management (UCM[1]) to describe best practice that linked change to the modifications made to the software assets—essentially creating the connections between entities on the left and right sides of Figure 3.4. To do so necessitated all changes, irrespectively of their size, be treated uniformly by establishing a register of change that must undergo a configurable review and approval process that offered the necessary governance and management oversight.

This relatively recent software change paradigm built on the already mature capability of earlier SCM tools that were designed to maintain versions of software artifacts as they changed—this feature made possible by software's digital manifestation. Version or source code control systems, as these early

---

1. Unified configuration management (UCM) is the term used to describe the manner in which ClearCase, IBM Rational's software configuration management tool, treats all development activities in a consistent manner so they can be branched, merged, integrated, and tested to become a software release.

forms of SCM tools were called, provided a full audit trail of all changes made and allowed the creation of baselines or snapshots to record the specific version of the artifacts that went to make up a specific configuration. These attempted to freeze in time the continual evolution of a dynamic software system, with a key baseline being the release (recognizing the ambiguity created by its use as a verb and a noun).

Methods similar to UCM are routinely applied by software developers today, utilizing a wide range of SCM tools that provide the enabling capabilities for many of the agile development practices. In the often volatile software industry, these tools help control and coordinate rapid changes by separating the management of changes from its actual implementation—essentially allowing the planning of releases (or change packages) before a new version of the solution is released to the customer.

As a continually evolving system, the enterprise has much in common with a software configuration, and SCM tools provide a model for the capabilities needed at the enterprise level. Indeed, Figure 1.1 is an enterprise variant of a simple version tree illustrating the branches and parallel streams that are common in software. The argument we make is that the lessons of software can apply equally in the complex, distributed, devolved, and interrelated systems and processes that comprise the enterprise.

But while the parallels are compelling, SCM only points the way to unraveling the management of interdependencies that remain somewhat problematic in software. The answer lies in the realization that the complexities inherent in a dynamic system can be tamed in the temporal dimension and utilizing familiar SCM constructs, such as the baseline and the release, can be powerful management tools when suitably scaled.

3.1.2.2 A Planning Scenario

While coordinating multiple changes and their impacts is a routine occurrence in software, the granularity of the problem and its very ubiquity does not expose its full management implications. For one thing, dependencies in software are not explicitly identified and are instead embedded in the source code or build scripts. This makes it impossible for management to have visibility of the interfaces and interdependencies. SCM and its baselines are limited to managing entire product configurations, as packages or collections, without knowledge of the internal design interdependencies. So the system-level dependencies are relegated to the technical domain and are not recognized as management issues. Yet many development challenges are clearly miniatures of the enterprise pattern that have their management and planning counterparts across the organization, and which can give rise to chaos when not appropriately addressed.

The coupling or interference that arises in planning and executing even relatively simple initiatives can be illustrated by the scenario presented below.

The emergence of somewhat counterintuitive, nonlinear system behaviors that arise bear witness to the need to consider even a small number of interdependent changes holistically. Implications of the dependencies in the enterprise pattern are visible in the planning, or temporal, dependencies that are shown to require significantly more coordination and collaboration (or negotiation) to achieve an agreeable release of an evolving system.

These dependencies are rarely managed explicitly; instead they are assumed to be understood and often buried in plans and designs that are not always communicated to those making the individual changes. The wider context is easily lost when the focus shifts to execution where the individual requirements merge with those from other initiatives and lose their original purpose. The tendency is to focus more on identification and allocation of the individual tasks, allowing the traceability to the original scope to be lost. Even when requirement traceability techniques are employed to retain the link to the original higher purpose, multiple tasks make for a jumble of interdependencies. The traceability matrix is completed retrospectively at the latter stages of execution and its detail more suited to post-development review and audit. A shared, high-level perspective of these dependencies is what is needed to guide an initiative from the outset and ensure those executing the requirements understand the complete context.

Consider the example, based on a real-world experience, a customer billing system that comprises multiple IT system components, as shown in Figure 3.5. In this example, the focus is on a small subset of the portfolio, where a high-profile business change requires a more elaborate customer bill format which displays a variable "customer greeting" that could comprise marketing information or a seasonal greeting.

In this scenario, achieving this new capability is seen by the business as a critical competitive feature given its strategy to improve customer loyalty through consolidated billing and communications. Yet to implement this

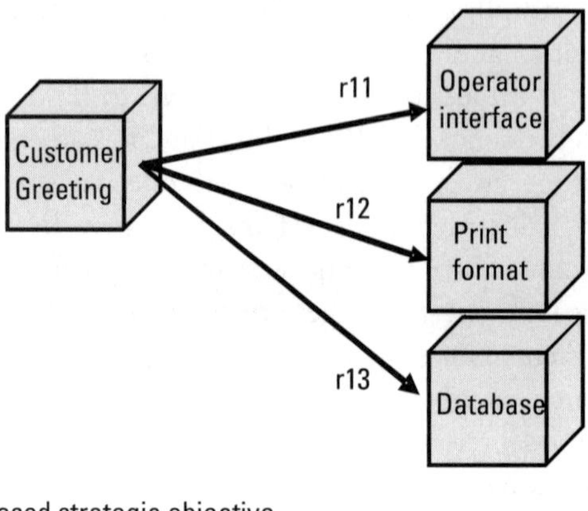

Figure 3.5 Decomposed strategic objective.

relatively simple requirement means that coordinated changes have to be made to several components of the billing system. Changes are needed to the bill print format to accommodate the new message, while the operator interface requires the addition of screens for the entry, review, and selection of the customer message for each bill run. Finally, the database tables need modification with the addition of new fields to store the available messages. Obviously, changing just one component alone would not suffice since all three capabilities (and so components) have to be released together to achieve the desired business result.

Most important, for understanding the enterprise pattern, the above change is to be implemented in the same timeframe as a separate, operationally important, but low business priority need to upgrade the printing equipment shown in Figure 3.6. This apparently isolated and routine IT upgrade however can come into conflict with the highly visible business initiative because the print formatting capability needs minor modification to support the new equipment. The challenge that this second change introduces is the parallel development scenario familiar to most software developers. It is a pattern itself, capturing the case where two changes implemented concurrently by different developers run the risk of having one change overwrite the other because of the lack of visibility of each other's actions.

Concurrent activity, or parallel development, is a classic SCM problem that is addressed by the more sophisticated tools which essentially aim to raise the visibility of multiple changes that impact the same software artifact. But, as will be discussed in the next section, while there are important lessons SCM can offer for managing a change portfolio, the enterprise scale introduces new and distinct conditions that need to be considered.

In terms of the example, planning the two changes gives rise to problems where resources are stretched. These relate to the temporal, or planning, issues that become evident whenever changes impact common assets or shared

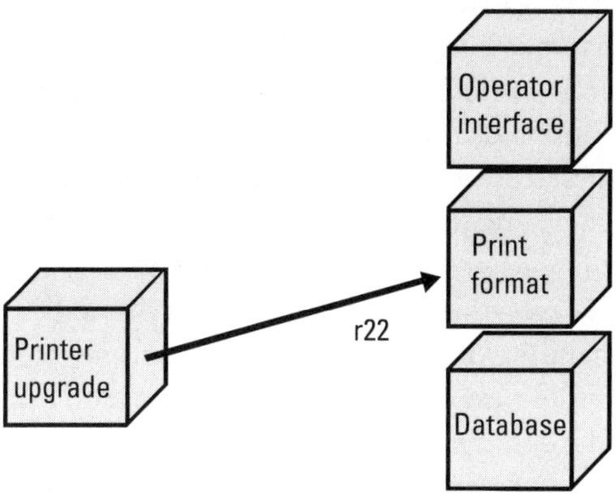

Figure 3.6 Apparently isolated change.

resources. The possibility of conflicting demands and priorities need to be managed and here the question of which requirement the print format team should work on first becomes the issue. Irrespective of the business urgency, it may be that the longer development timescales needed to change the operator interface and greeting storage makes it pointless for the print component team to schedule the corresponding formatting changes with the same priority.

In theory, all the dependencies described could be captured in a single "master plan," which could link the twin changes needed from the print format team by a finish-to-finish lag. With resources assigned, a critical path method calculation would then be able to represent the difference between the new message request and the scheduled finish of the other, slower tasks as float. But such a comprehensive plan presumes that both changes were planned together and all dependencies were known. This is more feasible in a carefully planned program, which serves to illustrate the key point that, in the enterprise, there is no planning done across the portfolio initiatives. The separation of responsibility for the different project and other change initiatives result in siloed planning. The creation of consolidated plan for the enterprise is rare and exceedingly difficult. It is not that such a plan for all changes in the enterprise portfolio cannot be created; it could with sufficient effort. However, there is rarely enough visibility to start such planning and attempts at too much rigor are likely doomed from the start. What is needed and motivates ERM are more agile, lightweight planning techniques designed to manage a disparate change portfolio.

As each change initiative emanates from different parts of the enterprise, it is more likely that those sponsoring each change manage them separately. In each of the figures above, the implications of the dependencies on the print format team may not even be known. It becomes the print format team's problem to meet the twin requests, and in all likelihood, schedule these changes together with other requirements as best they can. Given the available float in this scenario, there are not the difficult prioritization issues that could arise if demand exceeded their delivery capacity. However, in both instances the priority of the business change may not be effectively communicated to the executing team. Difficulties interpreting priorities arise because the delivery of the strategic goal cannot be easily translated into the more granular tasks involved. Even when some tasks are completed early, the business value—in our example the ability to flexibly communicate with customers on their bills—can only be delivered when all three separate tasks shown in Figure 3.5 occur.

The risk is that everything can become a priority, and teams perform tasks without an understanding of their effectiveness in achieving the end goals. The print format team in our example could unwittingly duplicate effort, changing the bill layout for the current print equipment and then again when the printer upgrade is complete. Separately, the operator interface might be implemented as a priority, in advance of other competing interface changes, only to find that it cannot be released because the supporting database and print format changes had not been completed. Clearly the priority of the outcome does not mean that the allocated requirements should be implemented with the same urgency. Such simplistic assumptions can lead to suboptimal use of limited resources—in rushing to deliver we become inefficient and sometimes ineffective. Yet the pressure to deliver, combined with the lack of oversight across the different initiatives, makes it impossible to create a nuanced plan.

3.1.2.3 Planning the Release

While the change impacts and corresponding dependencies arise due to the reliance on core architectural components (in this trivial instance, the shared printing equipment) the planning alternatives that result are about the time-ordering of tasks around these constraints. The problem is one of sequencing and prioritizing the different change activities so that they satisfy the business needs while at the same time offering a stable target state for the enterprise. Balancing the demands of multiple customer demands with the architectural reality defines the management challenge associated with the enterprise pattern.

Planning the configuration of the target state for a release requires consideration of the different combination of changes and the alignment of their delivery schedules. Release planning needs to work across multiple potential target states, each of which may become operational given a specific set of events. In reality, juggling the different possibilities and their consequences can make detailed planning impractical in rapidly changing environments—planning takes longer than the time-constant for change. Multiple dependencies make determining an appropriate schedule difficult, so it often easier to simplify things by constraining the temporal dimension and establishing fixed release milestones. These timeboxed deliveries are common in software and, commonly termed Sprints in line with Scrum terminology, support an iterative, often regular, release cycle.

Even in our simple scenario with just a pair of changes, illustrated in Figure 3.7, the full implications of the dependencies are difficult to capture or understand without full visibility across all the separate component teams. What might appear as distinct and independent changes can be found to be in fact coupled as a result of implicit planning dependencies. Adjusting the schedule for one change will impact the other, resulting in an explosion of potential temporal configurations. And in the real world, the dependencies are not likely to be as simple and obvious as in our example. There will be in all likelihood, other changes to the system being undertaken concurrently that could impact to other parts to the customer billing system, which are represented by the ominous, empty boxes shown in Figure 3.7.

If simply planning two changes with one common impact can give rise to the many issues and implications considered, what happens when a more

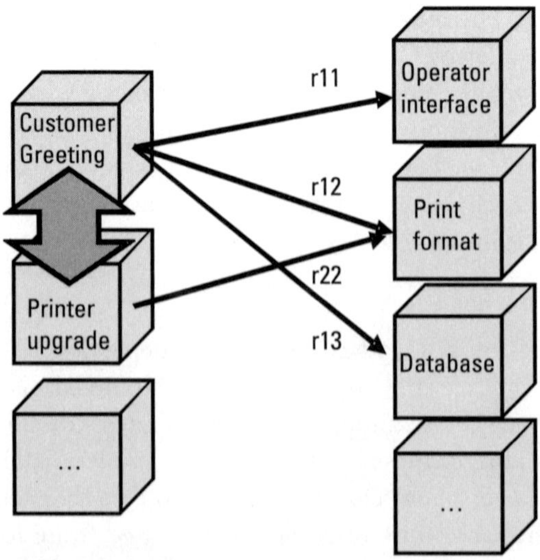

Figure 3.7 Implicit planning dependencies.

realistic scenario with even more interdependencies is encountered? The planning options compound if, for example, the printer upgrade also required a modification to the operator interface. And as we are considering not simply the software architecture but the enterprise architecture, there would be the need to consider the communications, operational processes, and training that may need to accompany each of the interim solutions that may be released even where they might be temporary. Further analysis of the requirements might discover that other business sponsors are unwilling to have the printer upgraded if it were to be before end-of-year reports are distributed. Such constraints emerge not as a result of the implementation of the changes themselves but from their collective behavior—the result of the interactions (or interference) between changes that involve the enterprise's capacity to both implement and absorb the entire change portfolio.

As this simple release planning scenario makes clear, careful release planning is needed to cater to implicit planning dependencies between what may appear to be unrelated changes. The implementation plan for the customer greeting is now intimately bound up in the planning for the printer upgrade—what seemed separate changes are in fact related by their shared dependence on the print format component. Indeed the scenario highlights that a network of dependencies (those empty boxes in Figure 3.7), while not always recognized or managed explicitly, links different initiatives together. The sharing of components and resources have the effect of coupling changes together and give rise to system behaviors that cause even small schedule adjustments to have knock-on effects. Delays in one area ripple through the system in ways that are not easy to assess. The result can be that a complete replanning exercise is needed for even small adjustments to a plan, requiring tremendous planning effort that can be out of proportion to the actual development required. It is this sensitivity to minor and indirect perturbations that creates management uncertainty and makes the enterprise pattern unsuited to the traditional, plan-driven disciplines that are, arguably, not designed for and are therefore inappropriate for such situations.

Having a comprehensive plan that captures the dependencies between all change initiatives in the change portfolio is time-consuming and can give rise to considerable management overheads. Attempting to represent such detail using today's project management techniques and tools quickly becomes incomprehensible and difficult to maintain. Changing priorities and balancing competing concerns across multiple sponsors, stakeholders, and component teams is what differentiates the release planning in the enterprise from more conventional project management. The implicit dependencies inherent in the enterprise pattern make planning changes in isolation ineffectual. Instead, a more holistic approach to release planning and management demands greater collaboration and communications.

3.2 Enterprise Release Planning

The granularity of the change initiatives discussed in the example above are purposefully ambiguous. As described, they could be managed either as projects or change or enhancement requests with the implementation ranging from changes to a few lines of code to the upgrade of entire IT systems. When a change is small enough, it might be one assigned to a single developer while if it impacts one or more IT systems, it may be managed as a project with execution responsibility spread across these separate IT teams. The scalability of the enterprise pattern provides a powerful motivation for managing changes of different granularity and type in a consistent manner across the organization.

A common scenario is to have changes of different granularity being managed differently with more significant changes being implement through projects while more modest changes become maintenance or enhancement requests. Organizations often separate the processing of these initiatives based on their source in the enterprise, scope, budget, or the nature of the artifact impacted (see the discussion in Chapter 5). Yet, it is not uncommon to find dependencies between these changes that are only identified late in the release cycle. Processing change by such arbitrary criteria does not allow visibility of impacts and can be inefficient with multiple governance bodies spread across the organization. Indeed, it was the author's experience of a similar situation that was the key impetus for studying the management of such dependencies more closely.

To effectively plan a release that encompasses multiple interdependent changes of varying scale and priorities is a systems problem that introduces the need to harness disparate changes in the enterprise. The challenge of release planning is that it goes beyond static, architectural interfaces and must instead consider the temporal dimension—the alternative transition architectures that arise as multiple changes have to be implemented. Each potential target-state to be released is but one point in a complex multidimensional universe of potential change (or version) combinations. From these numerous planning options, and their dependencies, management needs to make the decision as to what presents a valuable, and viable, baseline of the enterprise. For reasons of expediency all possible combinations are not able to be considered by decision makers—indeed, in the rush to deliver, the planning at this level can be ignored. It is from this management blind spot that unexpected and unintended consequences, usually unpleasant, can emerge.

3.2.1 Planning as Temporal Simplification

Planning a release is designed to simplify execution and delivery. It does this by selecting the changes to be implemented and determining the combination of component versions that will need to be delivered. In order to simplify dependency management, the changes included in a planned release should be selected based on their relatedness or cohesion and not simply on business priorities.

In an echo of the change and complexity discussion of the previous chapter, the enterprise pattern treats process and product (or activity and architecture) as two separate dimensions. Managing the former involves familiar reductionist techniques like the work breakdown structure (WBS) that decompose the scope and so risk sacrificing the holistic system perspective. To reduce complexity, the same principles applied to the product dimension decompose the architecture into smaller, discrete components. Each of these reductionist approaches have their adherents, which can simplistically be grouped as project managers and architects. The former consider the logical relationships between project activities while the latter focus on the interfaces between the static architecture shown previously in Figure 3.3. However, the relationships between the two dimensions is rich with management information but can be overlooked in our tendency to drive specialization that leaves us managing in silos.

Release management, as discussed in Chapter 1, can provide a bridge between the process and product dimensions. Purely reductionist logic leaves us with the challenge of synthesis and integration, which release management addresses in the project life cycle. When the problem scales to multiple project and architectural components, a more generalized form of release management is needed. This must bring together the twin pillars of management in the large: project portfolios and architecture. In the enterprise context, release management has to play out in the dynamic, multidimensional space that describes all possible future states of an evolving enterprise. The enterprise pattern can help us better understand the planning and execution challenge involved in achieving ERM. Figure 3.8 highlights the change dependencies that represent the impact on multiple changes on a decomposed enterprise architecture. This is the management gap or blind spot that exists in many organizations. It arises because decomposition techniques, while making a problem easier to manage, can create artificial boundaries. And our current management practices have inadvertently created a divide, separating business change from architectural complexity, which makes it difficult to bridge the two at an enterprise level. However, failing to do so raises the risk that the enterprise becomes fractured, with vital operational, system-level behaviors adversely affected. Decoupling the management of change from complexity presents a self-imposed business risk and makes it necessary to establish higher-order, more holistic governance at the enterprise release level.

Therein lies the integration challenge that is the natural consequence of decomposition and modularity. Planning an enterprise release forces a determination of the target configuration, the to-be state of the enterprise, and attempts to simplify execution by bundling changes with common impacts.The complex enterprise system comes with equally complex instructions about which

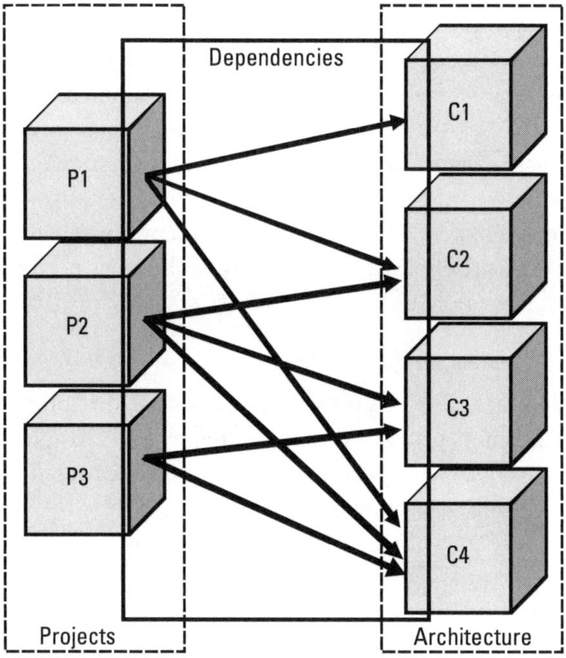

Figure 3.8 Managing interdependencies.

configurations are both allowable and support a viable business model. Of all the myriad random configurations possible for reconstituting the enterprise from its modular components, identifying the appropriate configuration to be released requires an understanding of the interdependencies between components. Yet few organizations have visibility of what these dependencies may be, as they are managed separately or, at best, only understood in the context of an individual project's solution architecture. And while planning the system evolution becomes essential to ensure the appropriate integration of enterprise system, it remains an anathema to agility. This then is the dilemma management faces and represents the real-world challenges presented by an enterprise system and the management gap found in the fourth quadrant discussed in Chapter 2.

The systems engineering approach taken in software disciplines such as SCM and software release planning offer a counterbalance to reductionist thinking and the challenges of system decomposition. These disciplines attempt to retain the integrity and continuity of the system and address changes and their integration in the temporal dimension. SCM tools and their version control capabilities have been integral to balancing these dual concerns. By maintaining detailed records of each individual change to software artifacts, a rich catalog of branching (and merging) patterns [1] could be supported. In this manner, SCM maintains the systems perspective by exploiting the temporal dimension; where things are decomposed by reductionist thinking they can be reassembled over time, through a series of releases, to reestablish the whole.

Software development therefore can offer fresh insights where the apparently contradictory objectives of rapid change and solution integrity need to be balanced. Software systems need to maintain a coherent whole in order to provide the capabilities—it is recognizably a system. And so it is in the enterprise with the increasing reliance on software-based IT systems that could support many alternative business strategies and processes. This suggests that the role of release management in software needs an enterprise equivalent. But the corresponding enterprise release management techniques need to ratchet up to provide value in the whole of business context.

3.2.1.1 Beyond the Software Challenge

For software developers, SCM provides an essential discipline to resolve the temporal, concurrency problems encountered when changing applications. Some useful techniques have been developed to balance rapid change and the need for system stability. Version control tools record each completed modification made by developers in a shared repository that supports what is termed "a single source of truth." With all changes made to artifacts in the repository, future merging and integration of these changes is made possible for separate, potentially parallel, releases of the software. Building on the repository, features termed change-sets attempt to logically separate the implementation of a change from the point at which it is implemented.

However, while SCM can offer significant assistance to developers, especially in larger distributed software teams, its ability to solve the more general case of the enterprise pattern is limited. The first and obvious thing that limits SCM is what makes it feasible in the first place. That is the medium of software, which (source files being primarily text) makes it amenable to software manipulation and that makes these tools viable. Version control capabilities allow deltas, or differences between source files to identified and several separate change (branched) versions to be viewed and merged back together. Such tools, combined with the ease of making changes to software, mean that it is possible to implement multiple changes, keep all these versions, and provide automated support to merge them together (albeit with varying degrees of integration effort) at some future point.

While powerful in the software domain, equivalent, SCM-like features are not possible to implement in the enterprise context. The difficulty of making and manipulating changes to the other dimensions of the enterprise architecture means that anything not completely software is inherently less agile—even where similar tools like CMDBs exist for managing networked hardware configurations and continue to grow in capabilities. SCM and the tools that support it come into their own after the source code changes have been implemented, allowing discrete changes to be combined and integrated into alternative release configurations. The semiautomated merging of discrete changes underpins the latest aspiration of continuously building a software product after each change. Importantly, this is only possible because of another clear distinction that software has: its ability to withdraw or undo changes if they prove to be flawed or inappropriate. Decision reversal in an enterprise context is likely to be far more painful.

So while the ease of making and removing changes in software can allow for greater agility in the implementation of changes, EA components lacking the flexibility of software have to undertake greater analysis and planning prior to initiating the change. The communications and negotiations entailed in release planning can be the antithesis of agility. Lengthy deliberations and slow decision making are expensive, burning time and resources—making it easy to understand the frustration of software developers, especially when it can often take less effort to implement all possible options than to undertake the reviews and analysis needed to gain management approval to commence work.

In general then, where dependencies are present as in the enterprise pattern, release planning necessarily has to consider the alternative combinations and options for implementing changes that result in a viable release. This is similar to software release planning in that it is about considering possible future release configurations, but without the benefits of software's innate flexibility and the capabilities of SCM tools. Enterprise release planning needs to be conducted more rigorously since the enterprise is less flexible, unable to compensate for poor planning. Small adjustments may be possible but, like steering an ocean liner, the enterprise moves at a slower, more stately pace that requires more up-front deliberation and planning.

Therefore, while software has its unique features that allow it to embrace change, the enterprise challenge is significantly greater. Instead of a software architecture needing to be integrated, tested, and released, there is an enterprise architecture to be grappled with. Planning alternative future states of the enterprise entail the additional dimensions of EA, including strategy, organizational structure, business processes, as well as IT and non-IT infrastructure. The diagrams representing the enterprise pattern could be made to appear a great deal more complex if components from these dimensions were to be included.

Clearly, some simplification is warranted to manage the moving target that is the modern enterprise. While software can provide some pointers, the differences between software architecture and an enterprise architecture come down to the differences between technical complexity and management complexity. While technical complexity can be hidden or delegated to specialists, in the enterprise everyone is a party to, and painfully aware of, management complexity.

3.2.1.2 Technical Versus Management Complexity

The challenges represented in the enterprise pattern span several different dimensions. The multiplicity of change initiatives is one and the modularization of business architecture is another, and both of these dimensions play out in time. But while the pattern represents the execution and delivery challenge, it can apparently be managed to some extent, or we would not have any examples of successful enterprises. However, the scale at which the pattern appears directly affects our ability to address the associated management challenges.

Complex engineering systems can be expected to have significant interfaces and interdependencies and yet have been successfully created and managed. Indeed, componentized architectures have been a boon to innovation and have resulted in a dizzying array of new, cheap product offerings. So it cannot simply be that the growth in componentization or even the increasing prevalence of change is what results in the problems encountered in the enterprise.

It is not the technical complexity represented by the enterprise pattern that is its distinguishing feature—if it were, humanity would not have progressed much beyond the mechanical watch. Engineering management disciplines have been addressing such technical complexity for centuries. Rather it is the packaging of complexity that creates the management problem; the inability to contain it within either an individual's capacity, or in the example of agile methods, within the bounds of a small, cohesive team.

The enterprise pattern becomes more evident when it bursts out of its traditional "technical" confinement and instead becomes a shared problem involving multiple stakeholders across the enterprise. An example of different parceling of essentially the same level of complexity is presented in Figure 3.9. In the first instance, all changes, components, and the interdependencies are kept within a single management responsibility. That responsibility may be that of an individual, team, or even a large organization—it is the span management control that is the key point. In the second instance shown in Figure 3.9, the responsibilities for both change and architectural components are assigned differently across the organization. In this instance, the same level of technical complexity and interdependency now requires communication and coordination across several different groups or teams. It is crossing different organizational boundaries that give rise to many of the challenges in the enterprise, with business modularity giving rise to increasingly federated structures that add to the management complexity. This applies to both the responsibility boundaries for the EA components as well as where in the organization changes emanate from.

It is these organizational aspects of change and complexity that differentiate ERM from the technical disciplines of CCM and SCM. Agility faces significant obstacles where dependencies can no longer be hidden within a small team or even an individual's competency. Instead, multiple parties have to be involved in the change process and the need for a coordinating and facilitating role emerges. That is, the case for management becomes more pronounced.

Increasing modularization and specialization means that the enterprise pattern is most evident not from a system or enterprise architecture perspective, but through the organizational structures that manage them.

The previously discussed planning dependencies now have to be conducted in full view of the leadership (who are in fact drawn in to adjudicate) and

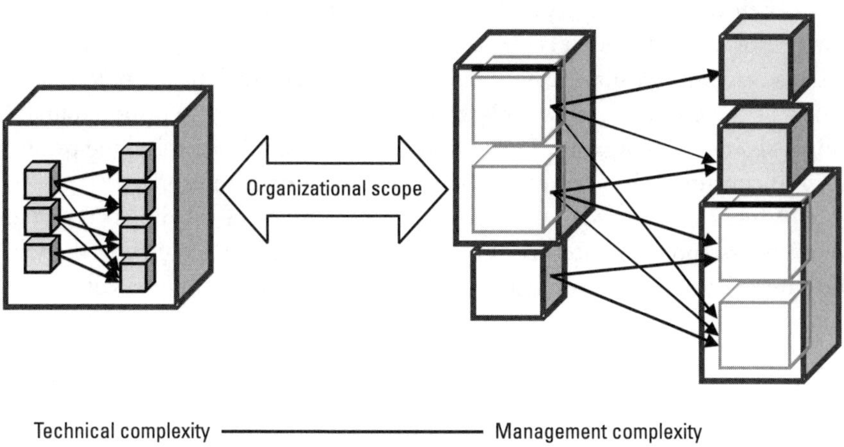

Figure 3.9 Technical versus management complexity.

subject to the differing agendas, priorities, and business strategies that make consensus virtually impossible and trade-offs difficult. Where management decision making has to be distributed across the organization, it is only the common purpose of the enterprise—its strategy—that offers cohesion and can create a shared focus. The question that needs to be addressed is whether current management thinking, more suited to technical complexity, can adequately address these new challenges and deliver the strategic change in the enterprise.

3.2.2 Enterprise Cohesion

While the word enterprise is commonplace and overused, there is little discussion about what that means and how an organization becomes an enterprise. Is it simply a matter of scale or are there other factors that give rise to the richness and complexity that distinguishes the enterprise? If an enterprise behaves like a dynamic system with multiple dependencies between components, it is worth looking at how this manifests in the organization.

Increasing the scale of the enterprise pattern sees the relationships between changes, and indeed business functions themselves, become more distributed and devolved. With increasing specialization in businesses, the pattern comes to describe more than simply a system, with a complex, modular architecture causing the enterprise to behave more like a system of systems.

While this comes close to exposing the essence of a large and federated enterprise, it raises the question of what creates the cohesion. What is the nature of the forces that bind the enterprise together in the first place? Clearly, at some point, an enterprise becomes more than the sum of its separate components. The emergent properties that arise from the enterprise system have to represent some commercial or competitive advantage, otherwise the enterprise would simply disintegrate into separate businesses.

The extent of mergers and acquisitions in the economy attest to the advantages of scale whether it is for vertical or horizontal integration. We first look at the advantages within the boundaries of an enterprise. Finally, we look more widely at the economic marketplace and how value chains, and more tangibly supply chains, span organizational boundaries and geographies, linking producers, and consumers in what might be termed the virtual enterprise.

While these considerations go beyond the scope of this book, they help us to understand how enterprise cohesion is manifested. Recognizing the form the enterprise pattern takes helps establish where ERM techniques can make a contribution.

3.2.2.1 Binding the Enterprise

So what is the form of the complexity in the enterprise that makes for its unique management challenge? What gives rise to the enterprise system? While these questions may seem pointless to some, given the phenomenological evidence, they go to the heart of why so many management endeavors prove ineffectual in reality.

Viewed from a systems or IT perspective, tangible features that bind an organization together are the interfaces and dependencies between the business components—its architecture. Yet, the wider implications of the pattern can be recognized from what is known as Conway's law, which suggests that IT architectures can be seen as a reflection of a deeper business reality. Melvin Conway in 1968 stated: "Organizations which design systems...are constrained to produce designs which are copies of the communication structures of these organizations."

While it is possible that siloed business units exist as completely separate divisions in the enterprise, in reality it is a rare, mostly temporary situation. Intertwined dependencies arise naturally because the better sharing of assets and specialized expertise across the organization can improve the bottom line. Economies of scale result from the improved utilization of expensive infrastructure and assets, which might be manufacturing plants but are increasingly IT related. So business efficiencies result from breaking down silos but the increased componentization comes at a management cost.

Strategies that target business efficiency often aim to centralize core business functions such as finance and HR. Such initiatives can reduce staffing overheads, standardize processes across the enterprise, and better share IT systems while decommissioning redundant systems. Centrally shared business services can then be deployed effectively and consistently across the enterprise, becoming a specialized center of excellence within the enterprise. At least that's the theory.

Figure 3.10 shows a modern-day enterprise that has three lines of business serviced by commonly shared functions: finance, human resources, customer relations, and logistics. We could also add other core services such as IT, product development, innovation and research. Any customer requiring a product or even a rudimentary service from any of the businesses has to interact with several departments in order to complete a transaction. For example, to fulfill a customer sales order might require dealings with the sales department for order processing, the finance department for billing and back-office inventory, and logistics systems for order fulfillment. Ideally, these interactions are transparent to the customer with IT enabled capabilities such as straight-through processing. But when processes are unwieldy, or break for whatever reason, the inner machinations of the business can become an impediment. Anyone who has been a customer of a large bureaucratic organization can relate to the difficulty in addressing even simple issues that cross internal departmental boundaries.

Customer interactions and other business processes span both business specific and shared services to thread the different business functions together,

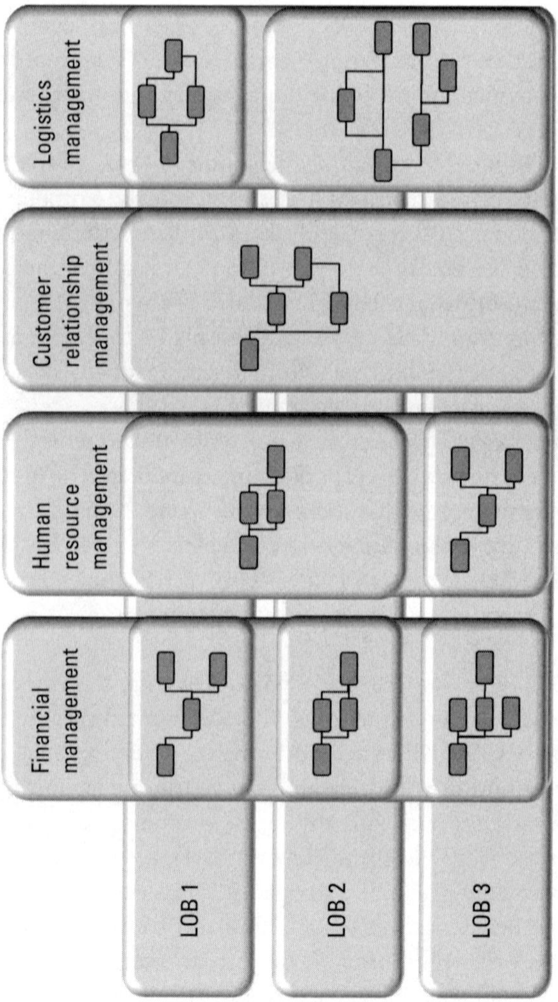

Figure 3.10 Cross-functional teams.

acting as the glue that knits the enterprise together. Sharing therefore results in cohesion and the cross-functional services shown in Figure 3.10 can be considered to be the glue that connects otherwise distinct business units together. It is the shared services that represent the core around which the business units operate.

The corresponding management challenge emerges from this sharing, arising from the need to coordinate the different departments to provide an efficient and integrated end-to-end process. Hence, the sharing and reuse of core business components across the enterprise come at a cost. As was evident from the earlier example of the shared printing capabilities, there are real management consequences to multiuse of core capabilities. Complex interdependencies result in the need to balance the concerns and priorities of the different stakeholder groups across the enterprise. Superior execution in specific departments is not enough and can be diminished by the poor performance of others. So a holistic system perspective becomes essential to optimize enterprise performance.

Of course, IT enterprise systems work across functional groups and have been embraced precisely to address these types of challenges. Essentially we use IT to shield us from the difficulties of coordinating the interactions between different parts of the enterprise. By automating business processes we essentially convert management complexity into technical complexity. Even if all variants of the process could be addressed by clever software that completely automates the business process, the hidden complexity will become evident, and reemerge, whenever there is a need to change that software. At the point of change it becomes necessary to understand all aspects of the process and engage those parts of the organization that are impacted—and so the presence of multiple changes gives rise to the enterprise pattern. Although the pattern can often appear to be the result of parallel changes to IT systems, it is in fact the changes to the underlying business processes (and therefore the enterprise architecture) that are the ultimate cause.

Increasingly, functions such as finance, CRM, and HR are essential to all manner of businesses, as demonstrated by the success of enterprise systems software. The next level of maturation of these functions is to become generic offerings, with standardized processes that are largely independent of the nature of the business. When these generic services are combined with the globalization made possible by the Internet and broadband telecommunications, businesses providing such specialized support can more effectively partner with each other to create a marketplace that blurs organizational boundaries.

3.2.2.2 The Extended Enterprise

As businesses begin to define and componentize their processes and architecture, it becomes easier to find alternative, more cost-effective means to perform its core services. With much of a business' back-office functions now operated by commoditized packaged solutions, businesses can find several vendors competing to offer their enterprise systems. The outsourcing of entire business processes and functions is the next logical step already evident. Routine functions, such as processing of credit card transactions, are now commonly undertaken by specialized, external agencies which, while distinct entities, interface into the enterprise's accounting systems. As identified at the peak of the dot-com era, enterprise boundaries are becoming harder to distinguish [3]:

> For information systems, it is increasingly difficult to draw a line around an application system and say you own and control it. For example, as value chains extend beyond enterprise, supplier and customer systems become part of each other's information architectures.

IT outsourcing and offshoring is just the first wave of the gradual virtualization of the enterprise. This trend towards componentization and specialization is the ultimate scaling of the enterprise pattern beyond a single enterprise, and blurring into a wider, economic marketplace. Here the pattern captures the dependencies between separate organizations, which cooperate to deliver some market need. The producers and consumers in the pattern form an ecosystem that is made viable due to the benefits arising from that value chain they service.

What is pertinent to an enterprise operating in such a marketplace environment is the added difficulty of coordinating the delivery and assessing the impacts of changes. The participants in such a value chain can be expected to different legal entities that operate under formal contracts that define the products and services that they need to exchange. As a result, making changes to any part of the (eco) system requires a complex set of negotiations, that when agreement is reached, requires the implementation and testing to be undertaken across several, distributed organizations utilizing different IT platforms, software environments, and development processes.

Porter's value chain, illustrated in Figure 3.11, identifies an organization's strategically relevant activities in terms of its support activities or core services (horizontal) and the primary business activities. The extended, or virtual enterprise could operate in an environment where any one of these activities can be conducted by separate businesses. But what is needed is for there to be sufficient economic value to outweigh the overheads of managing the operational dependencies, otherwise the enterprise has the risk of "disintegrating" and destroying the value chain.

The scale of the discussion, considering changes a value chain and ultimately the market, appears to be bordering on economics. Yet it is the emerging reality, with enterprise systems growing and morphing into even larger systems of systems. In this brave new world of enterprise components, the focus shifts

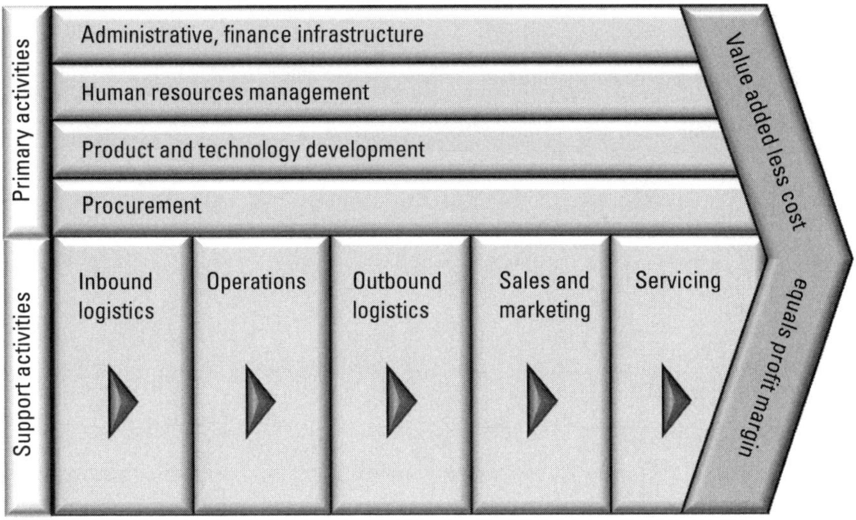

Figure 3.11 Porter's value chain diagram.

from the development, change, or even the operation of each component—they can be treated as a black box that is someone else's concern—to the complex set of interfaces and dependencies that give rise to the integrated whole.

The question that needs to be asked is whether our current set of management tools are capable of addressing these new challenges and delivering strategic changes under the conditions described.

References

[1] Berczuk, S. P., B. Appleton, and K. Brown, *Software Configuration Management Patterns: Effective Teamwork, Practical Integration,* Addison-Wesley, 2003.

[2] Ackoff, R., and Emery, F. E., *On Purposeful Systems,* Aldine-Atherton: Chicago 1972.

[3] Hasselbring, W., "Information System Integration," *Communications of the ACM,* Volume 43, Issue 6, June 2000.

4

The Project Paradigm

People in any organization are always attached to the obsolete—the things that should have worked but did not, the things that once were productive and no longer are.
—Peter F. Drucker

Change is so pervasive in modern society that many businesses have adapted by organizing themselves for continual evolution and transformation. In attempting to be agile, aligned, and responsive to ever-changing business conditions, enterprises have embraced the project construct. Adapted from its origins as a temporary structure for one-off, greenfield endeavors that essentially start with a blank sheet of paper, the project has become the means of delivering a continual stream of new products, capabilities, and services. A project-based organizational culture sees the enterprise geared for continual change, allowing it to undertake multiple projects or programs across distributed and loosely federated lines of business.

As the dominant management paradigm for executing strategy, projects and the project management discipline are now widely accepted with few questioning the suitability of the approach in the enterprise context. Yet the very nature of projects has evolved and been adapted in subtle ways to cope with modern business conditions. Even as each project in the organization attempts to maintain focus on its individual goals and defined outcomes, a project is increasingly a part of a portfolio of change initiatives being undertaken at any given time, each independently managed and funded by business unit sponsors. And while a project's objectives and management may appear independent, as discussed in the last chapter, they all seek to change a complex enterprise architecture. As such, they frequently require common resources and impact shared functions and IT systems over which they can exert little control and for which the systematic implications are often little understood.

This chapter explores the scope and limitations of the project paradigm in the enterprise context. This is not to be confused with a methodological debate about how a project is best managed; instead the discussion represents a more fundamental challenge to the project paradigm itself—and foreshadows the conditions under which it must adapt or transition to a release-centric paradigm.

First, as the favored approach for delivering strategic change, the fundamentals of project management are reviewed, looking at how the PM framework has adapted over time to become the success it is today. Next, the organizational impacts and structures resulting from the adoption of a project-based culture are explored, and the challenges projects face in addressing increasing complexity and interdependency. Finally, we review how the project paradigm is being extended and the project management office (PMO) is often established to manage an enterprise change portfolio that comprises multiple concurrent projects.

4.1 The Project Construct

There is little question that the project as a management construct is a remarkable success; a common and well-understood paradigm seen as a means of ensuring organizational focus and delivering results. The ubiquitous project is a concept applied to all manner of initiatives, from marketing campaigns through to IT, projects are used to develop and deliver business value across a range of industries and domains.

As such, project management (PM) is a respected discipline that has captivated management thinking and not without reason as it has been instrumental in numerous remarkable achievements. Multidisciplinary projects range across several domains including engineering, construction, IT, and software applications have helped distill the generic management techniques that are core to the PM discipline and form its body of knowledge.

However, the success of projects, as a management construct for one-off, greenfield efforts, does not necessarily translate into success in the enterprise. The very nature and definition of a project has been adapting over recent times to accommodate its new interpretation as a vehicle for delivering change. Yet simply using the project paradigm whenever management focus is needed to achieve a desired outcome may not be appropriate.

This section reviews the subtle transition the project construct has made from the early definition to the newer interpretation as a vehicle for delivering change. The evolving definition of what constitutes a project indicates how projects have adapted to the enterprise environment and its preexisting as-is operational architecture. This, compounded with the increasing professionalism of the PM practice, have set in place a new dynamic that can have the unintended consequence of diluting the discipline and separating it from the details of execution and delivery. The modern PM's focus on the generic aspects of managing a project relies on new specialized roles such as business analysts, solution architects, and subject matter experts that can serve to distance management from the coal-face—something which agile methods aim to redress.

Considering or questioning the fundamentals of projects in the enterprise can appear to be heretical, but it can help us understand and determine the management approach best suited for a specific initiative. The form of management needed for initial product development versus that needed for the delivery of changes where an existing architecture exists can be expected to be distinctly different. Lowering execution risk and improving the odds of delivering strategic outcomes relies on being cognisant of these essential distinctions.

4.1.1 The Changing Project Definition

Projects are ubiquitous today and the word has entered common usage even where it can sometimes be used indiscriminately to describe any kind of focused activity. Everything tends to become or is talked of in terms of a project; however, a project is more than simply a packaging of necessary activities.

The definition of a project has been changing subtlety over recent times; from its engineering and product development origins to the enterprise context where it is a vehicle for delivering change. With this expansion of project management discipline's domain or realm (in a meta sense, an expansion of the project's scope) has come an increasingly generic discipline; managing resources, schedules, and finances, but increasingly separated from responsibility for delivery of the desired outcome by a set of best practices that places the responsibility for results on others. It is in this purist, plan-driven interpretation of projects that there appears a yawning divide between responsibility and accountability, where traditional project management finds itself targeted by agile methods and needing to adapt.

4.1.1.1 Evolving Projects and Definitions

The Project Management Institute (PMI) is the world's largest not-for-profit professional association whose goal is to "advance the practice, science and profession of project management." The PMI's Body of Knowledge (PMBOK) defines a project as follows [2]:

> A project is a temporary endeavour, having a defined beginning and end (usually constrained by date, but can be by funding or deliverables),

> undertaken to meet particular goals and objectives, usually to bring about beneficial change or added value.

The above project definition is necessarily wide-ranging and represents the relevance of PPM terminology across different industries and a range of disciplines or domains. The definition covers the wide range of initiatives that are given the designation of a project. It encompasses change and transformation projects in the enterprise, enhancement projects for existing products, as well as new product development and traditional engineering and construction projects.

Having to embrace all these types of projects has resulted in stretching and sometimes abusing the basis of the term, and this will be shown to occur most clearly in the enterprise context. Here projects are rarely greenfield, and one or more of the dimensions of the enterprise architecture have to be catered for and transitioned from the as-is state. In the enterprise, projects deliver change in one form or another. Even when a new IT system is developed or procured, its implementation requires organizational and process changes that have a crucial influence over project success.

Change in a wider context than the technological dimension is therefore a common element of projects that is modifying its very definition. The focus on change sees more general definitions of the project emerging with the following elegant variant [3]:

> A project as a discretely managed, time and resource constrained activity to effect a specific change to the status quo.

The organizational changes associated with projects requires a new transparency. The newer social networking capabilities promised by Enterprise 2.0 have their project equivalent in Project Management 2.0, which heralds a new era of communication and collaboration. These fundamental shifts in the nature of projects bear further investigation—a review of assumptions and constraints that define the project construct—in academic parlance, its theoretical underpinnings. Because, even as the nature of a project itself has changed, there are few corresponding advances made in the project management discipline.

4.1.1.2 Projects and Change

As the project construct has evolved, examples of projects now span the full range from one-off, greenfield developments to change programs that transform existing business. But a project is more likely to be successful and suffer fewer difficulties when it meets the traditional criteria for what constitutes a project; what has traditionally been described by as the triple constraint: delivering a defined scope within financial and schedule constraints. Earlier definitions of a project and the project management discipline have had a product or engineering focus and were less equivocal about the project's purpose. A project had more to do with the creation of a new product or system than with the need to change an entire organization to adopt it.

But the project as a vehicle to deliver change follows a logical arc that makes the link between the two inevitable. It starts with the blurring of new product development and the less glamorous maintenance phase. Influenced by the flexibility offered by software, now products could be developed incrementally, maturing over several iterations. But as well as being a core tenet of agile methods, iterative development breaks down the divide between development and maintenance, instead creating a continually evolving product, punctuated by short development cycles or sprints. The growing influence and adoption of agile methods has meant that even greenfield projects can be treated like change programs, where they undertake an iterative development cycle and deliver successive, incremental releases.

However, more significant global events were about to fundamentally change the nature of IT projects and project management at large. At the time software's iterative development was making an impact, greenfield projects were the norm, with the industry's immaturity requiring most systems or applications to be bespoke developments. With the advent of increasingly sophisticated packaged ERP solutions and the growing market of these commercial off-the-shelf (COTS) solutions, the "build versus buy" decision became a critical management decision of the late 20th century. Arguably, it was the Y2K problem that finally put the question to rest, as replacing IT systems became a matter of urgency.

A lack of foresight by early programmers meant that most software solutions developed in-house would encounter problems if operated beyond the year 2000 and had to be remediated. The cost and difficulty of changing these often poorly maintained systems, with little or no configuration management of the software source code, to accommodate four-digit date calculations was a daunting proposition. As a result, it was an opportune time for businesses to upgrade to a vendor supplied and supported ERP solution that often offered new capabilities, avoided the remediation costs, and mitigated the risk of similar future events. The Y2K bug, as it was known, helped make the decision to buy instead of build a near universal decision for large enterprises—this deadline was not one that could slip. Together with several legislative and compliance changes of that time, including the adoption of the Euro currency, the imposition of new goods and services taxes, and the compliance requirements of the Basel Accords on the banking industry, the need for comprehensive maintenance of enterprise systems was highlighted, which drove the business case for the acquisition of vendor supported solutions rather than new product development and its associated maintenance.

The aftermath of these global, industry-wide trends has meant that projects in the 21st century are fundamentally different to those of the past. Projects increasingly focus on changes to an existing system or enterprise architecture rather than simply the development of a new one. In this new era of project management, the enterprise has moved past the initial condition, going from nothing, a vacuum, to one with a complex, shared, and interdependent as-is architecture. Now projects succeed not when they offer a technically superior solution but when the solution is adopted by users and provides value to the business.

Projects conducted in the enterprise are constrained by the as-is architecture and necessitate they be managed like changes—sensitive to impacts on the organization's people, processes, and technologies, they place greater demands on project managers. Project execution has fewer complications and is lower-risk when it delivers new, independent products and remains separated and independent from ongoing business operations. When a project involves a stand-alone system or service fails, problems can be contained within the project's scope. The difficulty with change projects is that problems quickly become visible to stakeholders and risk, adversely impacting business operations. In the worst-case scenario, there could be damage to business reputation and potentially, the viability of the business. This attests to the critical role that projects play in the enterprise, and the increasing levels of professionalism demanded of project managers. But their responsibilities can only go as far as their scope allows—beyond mitigating the implementation and integration risk of their deliverables, there is the wider problem of maintaining a stable enterprise platform in the face of multiple projects.

4.1.2 The PM Practice

The popularity of the project construct has seen the community of PM professionals grow significantly in the last two decades, with established institutions formed to promote the discipline and support and certify individual practitioners. The Project Management Institute (PMI), founded in 1969, is the largest international body representing the profession with a global membership in excess of half a million. The PMI is by all accounts a mature, successful, and vibrant professional body active in the advocacy of project management. It has established several professional standards and is author of a definitive collection of best practices for project managers known as the PM Body of Knowledge, or the PM-BOK. Through this comprehensive resource, related standards, and certification programs, the PMI represents a unifying force for the PM discipline and has helped establish a hegemonic project culture. With projects as the presiding delivery paradigm, numerous industry bodies have endorsed the role of the professional project manager and helped establish common terminology, tools and techniques that are accepted across a wide range of domains and industries. The PMI is not alone in the promotion of project management and local variant can be found around the globe, most notably the United Kingdom's Office of Government Commerce and its Projects In Controlled Environments (PRINCE) method, which is widely favored in Europe. The ISO 1006 standard for quality management in projects has likewise had influence on the global PM community.

4.1.2.1 The Generic Project

It is clear that the modern definition of a project could encompass most business, if not human activity. Certainly, professional bodies like the PMI have a charter to promote project management and benefit from a generalized definition of a project as it extends its influence in a growing market. Rather than stressing distinctions within the field that could separate or divide the profession, it attempts to identify common ground. This is not a new situation and earlier, more mechanistic definitions of project management that focused on the triple constraints pointed to a discipline that stressed the abstraction of scope, costs, and schedule over the specifics of a project's outcomes.

With costs and schedules essentially requiring accepted, well-understood disciplines such as accounting and planning, it is only the scope of the project and the technologies utilized which require specific knowledge. With the adoption of a process-based approach to project management, such as the PRINCE method, the discipline remains independent of industry, domain, or product, considering only the generic phases and steps that define the project life cycle. As such, the PM discipline is founded on techniques and tools that aim to apply equally across the range of possible projects. And within reasonable bounds, one project can be treated much like another.

Although the nature of the requirements gives rise to the greatest variability across projects, PM techniques aim to address the fundamental issues that relate to defining and managing the scope. The need for domain knowledge is reduced by adopting that classic reductionist tool, the work breakdown structure (WBS), which seeks to decompose the scope, or more specifically the to-be architecture, into more manageable sized components so that responsibility and accountability for their execution can be assigned to individuals or supplier organizations. More fine-grained decomposition gives rise to the requirements, which detail the capabilities needed and which are then managed and traced as unique and sometimes contractual agreements between the customer and supplier that then form a part of the commercial and functional acceptance criteria. The clinical decomposition of scope can be most easily accomplished in product engineering organizations, whereas in the enterprise, more people-intensive projects, including those in IT, can suffer from uncertain, unclear, and sometimes contradictory requirements. In such environments, scope management becomes more a matter of stakeholder management, and clear communications, which is discussed further in Chapter 7.

Aside from a project's scope or requirements, the remaining dimensions are largely common across all projects. Managing the time or schedule dimension utilizing classic Gantt charting techniques supplemented by the PERT or network analysis are valuable in larger and more stable programs. Where there is volatility around the scope and requirements (these are effectively changes to the change that is the project itself), there is less value in detailed planning as there is often significant effort needed to maintain up to date plans. This is the situation that motivated the Agile Manifesto's preference for "responding to change over following a plan."

The tracking of project costs relate to relatively simple, but not to be underestimated bookkeeping and accounting practices. Budgetary considerations and the allocation of costs to capital expenditure (CAPEX) or operational expenditure (OPEX) accounts are important aspects in determining the volume of work that can be undertaken by the business. Effective financial management of the portfolio can ensure favorable tax treatment of the strategic change budget, and more sophisticated reporting including earned-value calculations can provide insight into the health of in-flight projects. For better or worse, the techniques and skills required to undertake financial accounting and analysis are portable across different industries and have little to do with the particular domain in which the project is being conducted.

Finally, a key activity in a project is organizing resources and managing people issues—including communicating, negotiating, justifying, and persuading project personnel, stakeholders, and sponsors. Especially in larger organizations, it is not uncommon for project managers to "manage up" the organization hierarchy while relying on project administrators or other assigned personnel to manage the details of the project. The results of the above is that the project increasingly becomes a generic construct for the management of temporary activity. And with it, the role of the project manager likewise becomes a necessary and respected generic professional responsible for delivering change—without fully knowing what that entails or how it relates to the strategy.

4.1.2.2 The Professional Project Manager

The professional project manager today is often a certified PM specialist who can, theoretically, work across a range of businesses, domains, and even industries. This is a relatively recent trend with earlier PMs' career paths requiring they have domain-specific expertise. In the recent past, PMs came up the ranks in the industry in which they worked and had an intimate knowledge of the technologies and other specific characteristics of their field. The role of PM, like other managerial positions, was assigned to those with sufficient technical expertise to gain the respect of the individuals that they had to manage, even where they may not have had the requisite people or project skills to handle the position.

The modern PM, however, is trained in the techniques and tools that comprise the profession's large and increasing body of knowledge and acts more like a consultant, with the potential to move between industries to apply the generic discipline of project management. What they lack now, however, is the depth of domain specific knowledge that makes for deeper understanding of the project requirements. As a result, the increasingly PM role can result in decoupling management from the practitioners in the trenches who undertake the project's work. While this separation between management (of which the PM is in reality the closest representative) is in many ways unavoidable, especially in multidisciplinary project where a PM cannot necessarily have a background in all aspects of the project, it is the very thing that Agilists abhor.

However, career PMs have emerged in the past with their distinctive coordinating role clearest in project organizations that used a matrix structure, pulling together multidisciplinary teams to work on projects, from specialist functional resource groups. Projects in the enterprise share a number of similarities with this earlier model, as professional PMs have to rely on a team of specialists from across the organization to execute the program of work. Each phase of the project requires its own expertise, from the business analysts who solicit and analyze the requirements of subject matter experts across the business, to the testers and release managers who ready the final solution for the business. Between these two ends of the life cycle reside an array of specialists, including those in IT who support the different software applications and hardware platforms that the business relies on. Likewise, business process analysts and organizational change specialists complement the IT resources and ensure the solution integrates into the business environment. These roles constitute the PMs cross-functional team of experts that provide the skills to execute the project's mission; inspired, motivated, or simply goaded by the PM to deliver the desired outcomes within cost and schedule constraints. While the PM may not be able to dive deep into the detail of the project's scope and functional requirements, the project team's ability to provide the necessary synthesis and expert advice to management, stakeholders, and sponsors becomes crucial.

Aside from the large number of techniques that the professional PM has to master today, success in the role remains to some extent defined by force of personality and a pragmatic and determined delivery focus. Those individuals with a natural inclination for marshalling resources and directing others are thought to have a natural predisposition to be managers; and when such a disposition is combined with a fierce results-orientation we have the making of the classic PM. Yet, the complexity of modern projects requires the reliance on a team of experts that swings the success determinant more towards effective communication and collaboration.

While the increasing professionalism of project management is not necessarily a negative trend, it can serve to separate the project manager from the details of the work needed to achieve the outcomes. This can defocus scope and requirements management in favor of the more administratively straightforward and well-understood aspects of costs and schedule management. Project management can become a black box where the external form of a project is managed carefully but the actual capabilities being delivered are not well understood and are too often disappointing. This concern can be seen as a reprise of the recurring agile versus plan-driven debate that argues the two sides of the need for expert, craftsman-like solution development versus more scalable, process-based development that in actuality is designed to reduce the level of skills and experienced necessary by dividing the process into smaller, specialized tasks.

Bringing together the different facets of a project is essential to ensure successful delivery, and that challenge becomes more pronounced with increasing specialization that can fragment the project, making it difficult to deliver a coherent solution. The challenge is compounded in the modern enterprise where projects need to work closely with those operating the current state of business processes and IT systems. The increasing professionalism of PM combined with the changing nature of the project to become a vehicle for delivering continual change means that projects are now an integral and entrenched feature of the modern enterprise. And apart from their scale, projects now share a lot of the characteristics previously attributed to implementing smaller, more granular changes.

4.2 Projects in the Enterprise

Projects are a highly successful generic management construct that have been embraced across all industries, disciplines, and cultures. The generality of project management has allowed it to become the dominant management paradigm, but not without some significant difficulties. Perhaps something starts to go awry when complex organizations structure themselves for constant change and projects become "business as usual" in the enterprise?

What is clear is that while the project construct, and so project management, is an unmitigated success, actual projects in the enterprise have a poor track record for delivery. For over a decade, The Standish Group has been reporting on the high failure rate of IT projects in their regular Chaos Reports. Instead of this diminishing the appetite for project management, it has instead mobilized the project management profession to collect best practices in the previously mentioned PM-BOK and PRINCE frameworks, while calling for an increased professionalism in the industry through certification programs.

The risk of increased professionalism and specialization is that it can swell the number of participants involved in delivering a project. More specialists directly involved in the project life cycle create a coordination and handoff challenge and serve to spread the responsibility for poor outcomes. The creation of centers of excellence, or practice groups, for the different disciplines involved often give rise to specialized resource pools shared across multiple projects. While well-meaning and beneficial in certain contexts, these trends lead to siloed discipline areas that are the opposite of the broad, multiskilled, and self-organizing teams desired by agile methods. Specialization effectively shrinks the responsibility of the expert practitioners, giving rise to a process-driven project management life cycle that has to engage the different specialist practice groups such as business analysts, solution architect, and testers, negotiating the availability of these equally generic and specialized resources.

The lack of clear management accountability in the enterprise context can result in a project's end-to-end execution life cycle being compromised, and some fundamental tenets of project management begin to be strained. In particular, the need to share specialist resources across multiple projects dilutes the project manager's control. This introduces the need for increased soft skills in project managers who have to be accomplished negotiators and politicians (or alternatively, strong personalities who can bully colleagues into submission) to achieve the results needed by the project.

In this section, we take a look at some of the common features of projects in the enterprise context and the challenges they face. For all the focus on management best practices, the success rate of projects, particularly in the IT domain where both change and complexity are present in abundance, remain disappointing. The low project success rates reported consistently over the years show that project management is not a solved problem. There remain difficulties in achieving project success even with the benefit of the increasing professionalism of the field and the benefit of over a half-century of accumulated knowledge. Even where the enterprises adopt PM best practices in attempting to meet the market's demand for innovation and rapid change, increasing business complexity has given us new challenges that question existing management approaches.

4.2.1 The Project Culture

Embracing change in the enterprise has created a project culture in which the organization is structured to accommodate a perpetual change cycle. A significant, parallel project organization often exists to execute the change portfolio and manage transitions to a continually evolving target operating model.

In multiproject environments, the enterprise has to absorb a veritable wall of concurrent change projects delivering new capabilities in a series of incremental releases. Adapting to these successive waves of change sees the project construct take on a contradictory position in the enterprise, a temporary construct that has become a persistent feature in the enterprise landscape.

4.2.1.1 Project Concurrency and Backlogs

A root cause of the ubiquity of projects in the enterprise is the aforementioned propensity for change resulting from software-based IT capabilities. Business' continual, evolutionary discovery of the possibilities presented by new technologies has created an apparently insatiable demand for IT solutions. But the inherent enterprise complexity combines with the limited supply of suitably skilled IT resources to slow the solution delivery. This has meant that, to date, demand for new capabilities has exceeded the capacity of IT departments and given rise to business frustration.

It is the pent up and unsatisfied demand of the different stakeholder groups across the enterprise from which the strategic change portfolio has to be determined. These stakeholders represent different constituencies, lines of business, and market segments that seek funds and resources to undertake their favored initiatives. The competition for limited resources essentially means that some business requirements will simply not be able to be acted on, and so the emphasis on portfolio management is on the prioritization and selection between different and competing business cases.

Where business demand is great and resources allow, the trend is to execute multiple projects in parallel to maximize throughput and minimize time to market. This can exacerbate the integration challenge and impact the coherence of the enterprise architecture. A manifestation of the change and complexity cycle previously described, the risk of concurrent change is that architectural integrity, if it ever existed, is difficult to maintain in the face of a barrage of separate changes.

As well as eroding the likelihood of well-architected solutions that effectively share enterprise resources, the business demand for change results in a requirements backlog; that collection of lower priority initiatives or requests that are not able to be satisfied within the current portfolio execution cycle. These requirements remain on the business wish list to be considered and prioritized in the next portfolio planning cycle. In this way, the perpetuation of projects can be anticipated, and where the portfolio is well-managed, can identify a pipeline of future initiatives that will set the level of demand for enterprise resources.

The visibility of these future demands allows the enterprise to prepare for a wave of predicted changes with resource or functional managers planning for this anticipated demand. And so the cycle of perpetual change continues.

4.2.1.2 Shared Resource Pools

The effect of specialist roles, like those of project manager and business analysts to name but two, combined with the consistent demand for these positions over time, has seen the need for these resources to be retained by the enterprise. In anticipation of future projects, these different specializations form a practice group that develops its skills in the particular methods, techniques, and tools that are applied by that discipline.

This expediency is an echo of the matrix organizations that were popular with project-based contractors that maintained the twin lines of management catering to their functional specialization and project-based responsibilities. Similar to matrix organizations, pool resources only generate revenues when assigned to projects, becoming the temporary home for personnel between projects. The key difference being that in the enterprise, the funds raised are internal accounting transactions rather than real income.

In enterprises facing continuing waves of projects, resource pools can be an effective means of sharing resources, with the potential benefit of creating centers of excellence that develop appropriate standards and disseminate best practices. However, the key point is that resource pools represent evidence, if it is needed, of the ubiquity of the project paradigm and the subtle, somewhat paradoxical, evolution that has taken place in the enterprise where change is commonly treated as business-as-usual, permanently enshrined in the organizations structure.

Managing the demand and supply of resources in the project pool is problematic as future resource needs are dependent on the level of funding and the approval of project proposals. Maintaining the pool at the right level or capacity now requires careful resource management and, essential for forward planning, visibility of the project pipeline.

When project funds are available and business changes result in the demand for resources exceeding supply, projects need to be prioritized to determine which get access to the limited resources. Where supply exceeds demand, those administering pool resources look to market their capabilities and find them work. Acting as ballast for the ebb and flow of projects' demands, resources pools also have the unintended consequence of limiting efficiency initiatives, with the goals of pool revenue growth and resource utilization acting as a brake against project cost reduction and efficiencies.

Maintaining utilization levels of personnel in the pool can result in resources being assigned to work across several projects. While this makes economic sense, it is in many ways an anathema and contrary to project objectives because an individual's focus cannot be guaranteed to be on a particular project. The inefficiencies of multitasking can, and do, result in projects being executed poorly and become more like business-as-usual, operational activities.

Ironically, by developing a project culture in the enterprise, the very point of having projects has been undermined, which is the ability to have dedicated focus on an activity! With people working across multiple projects and different projects in contention for resources, resource pools are expediencies that have created new problems. And there are other aberrations that have resulted from the overuse and uncritical application of the project management discipline. Often difficult to pin down or identify in the detail, the consequences of the project culture in the enterprise are easier to see at the industry level.

4.2.1.3 Professionalism and Proliferation

There are risks in the increased professionalism of project management and the application of best practices that may at first appear counterintuitive; the risk that professionalism does not necessarily translate into improved project success. Certainly there is an assumption that acquiring the skills related to project management and applying the necessary processes and techniques will improve the probability of success. However, aside from any direct evidence, there are also attendant dangers that come with adhering to the processes and techniques slavishly.

The first difficulty relates to the fact that a trained professional PM, while committed to a project's success, applies accepted practices that demand project stakeholders participate in the process and fulfill their obligations. Like any professional, the PM expects others to undertake their duties and provide the information necessary for the project to advance. This is nowhere more evident in the previously discussed need for sponsors and users to provide their requirements so that an appropriate solution can be developed. The assumption that key project stakeholders should be knowledgeable enough about the business to either know what they want or at least be able to articulate their needs when prompted by an analyst, is one that is at the heart of the agile versus plan-driven debate. Importantly, adherence to PM practices can make professionals detached from the outcomes, allowing them to use the process as a means of transferring risk and possibly blame for slow progress or incorrect requirements.

The reality is of course that a project is a partnership between the sponsor and those who are charged with delivering the desired outcomes. The above observation only serves as a warning that when project managers focus only on the process dimension, it leaves open the possibility that the customer gets what they ask for rather than what they might really need. Managing simply "by the book" results inevitably in a dull, dispassionate, and unsatisfactory project execution. What is needed is a greater focus on delivery of the solution, or outcome, that is the natural preserve of release management.

Another consequence of the increasing professionalism of project management, or possibly the greater availability of project managers, is the proliferation of the role—in a single project. While this may seem contradictory, there are a great many managers in even moderate-sized programs of work in the enterprise where there is evidently a need for several project managers to work across the different supplier groups. This is a peculiar condition that is particularly evident in IT programs, where the greater interdependency and complexity of systems and their suppliers results in several individuals fill the role of IT project manager (although different titles may be used) to mirror the (business) project manager's role within each IT system group.

There is no fixed pattern in the multiplicity of managers with several different titles, including IT delivery managers, business implementation or transition managers, release managers, and other variants who are deployed across an enterprise program. What is clear is that no longer can one expect a single project manager per project; more likely a different PM is assigned to each supplier interaction or system interface. Clearly, this proliferation of management is not an appropriate response to enterprise complexity. Being neither efficient nor effective, the decision to divide and conquer only serves to increase project costs substantially and runs the risk of creating confusion and dissonance. Having key interfaces delegated to separate managers results in greater reliance on accurate communications and yet more meetings to coordinate and align activities.

While the professionalism of PMs does not necessarily contribute to such inefficient practices, the greater availability of PMs must not seduce desperate sponsors into thinking that quantity can compensate for quality. If we are to accept that the capabilities of today's best project managers are equal to, if not greater than, the skills of past practitioners, then the difficulties of managing today's programs must arise from some fundamental, underlying complexities that were not evident in project of the past. The enterprise pattern has already shown where new challenges can arise for projects in the enterprise. In the subsequent chapters we look to address these challenges by adopting the release-centric approach that is at the heart of ERM.

References

[1] Drucker, P., *The Five Most Important Questions You Will Ever Ask About Your Organization,* New York: Jossey-Bass, 2008, p. 54.

[2] *A Guide to the Project Management Body of Knowledge,* 4th Edition, Project Management Institute, 2008.

[3] National Audit Office, *What is Project Delivery?,* http://www.nao.org.uk/practice_networks/project_delivery/what_is_project_delivery.aspx (accessed August, 2011).

5

Release-Centric Planning and Design

Adventure is just bad planning.
—Roald Amundsen, Arctic explorer

Projects are the dominant vehicle for delivering changes in the enterprise, so less attention has been paid to the growing integration and interdependencies that make them behave like a complex adaptive systems. The project paradigm is so entrenched in most enterprises that its language pervades our thinking about the delivery of change, whether strategic, tactical, or operational. From the point of the inception, it is common to express new initiatives in terms of a future project for which a business case has to be established. Decisions at the highest levels of management are routinely taken after reviewing and prioritizing *project* proposals—the notion of a project as the delivery vehicle is established top-down and baked into our thinking from the earliest point in its life cycle.

However, the challenges facing project management in the enterprise should force us to confront some uncomfortable realities and ask whether the project is the appropriate vehicle for delivering the continual stream of changes that is now common. The rise of the project and the emphasis on agility needs to be balanced by tools for managing the architectural complexity and interdependencies within the enterprise. For business to be innovative, and yet maintain operational continuity, requires a more nuanced approach to change, one that considers potential impacts and dependencies so that the delivery of the selected change portfolio meets strategic demands without corrupting the enterprise architecture.

ERM and its release-centric perspective of enterprise change offers an alternative to the traditional project paradigm. Adopting a release perspective raises the sights above individual projects to bring both changes *and* their impacts more clearly into focus. It also challenges the status quo, suggesting that in a complex, rapidly evolving environment, a consolidated release, rather than a multiplicity of projects, is the preferred vehicle to deliver change in the enterprise. The simple argument is that with the interdependencies evident in modern businesses, projects simply cannot be allowed to execute and deliver independently a wider enterprise perspective. Instead, the business need for change and transformation has to be reconciled with the practicalities of stable transition architectures that can assure continuity of the enterprise. Ideally, the necessary release planning is is done early in the execution cycle at a point where the strategy is still being developed and before the selection of specific projects to fund. As such, the easiest first step to ERM is the implementation of a release-centric approach that extends project portfolio management (PPM) to ensure it addresses enterprise change more systemically.

A release-centric approach consolidates changes to the enterprise tackling the problem of uncoordinated project initiatives at source. Portfolio management, where the mantra "doing the right things" attempts to separate that "what" from the "how," can often set business expectations without establishing a clear execution plan. It emphasises what can be considered program design and proposes packaging business changes into an execution portfolio that has a predefined release roadmap. The contents of these enterprise releases need to be planned from the outset, rather than simply being a delivery vehicle for a collection of independent projects. The latter situation, while problematic, represents the business reality without a release-centric approach to portfolio planning. While ERM tools and techniques apply equally to the release delivery phase and are covered in the next chapter, the greatest benefit arises when the change portfolio is managed at the planning and inception stages.

To truly achieve control over an enterprise's evolution, the entire change portfolio should be managed in a unified and consistent manner that avoids the unnecessary and arbitrary distinctions often made in prioritization. Irrespective of the source, granularity, or delivery vehicle, whether it is a strategic initiative or relatively small enhancement, all changes need to at least be "visible" and undergo a release planning and design process that identifies the next transitional states, or releases, for the enterprise.

This chapter therefore sets out to discuss the early phases of defining and planning the enterprise release. This is most effective when *all* proposed changes are uniform, applying a common, unified change governance model. Such an integrated view of the enterprise's change portfolio offers greater insight into dependencies between changes that serve to reduce execution risk. Selected change initiatives can then undergo release-centric planning that goes beyond simply prioritizing the changes, to establish optimal release packages that become part of the enterprise release roadmap.

5.1 Enterprise Change Governance

Separately managed, projects in the enterprise run the risk of compromising the enterprise system and exacerbating complexity. Siloed changes have been tolerated, and maybe encouraged, precisely because they drive changes through the enterprise with minimal resistance—but this is not always a desirable situation and unintended consequences can be damaging, causing business strategies to fail. Breaking the change-complexity cycle requires that both be addressed in the enterprise; ignoring, circumventing, or suppressing one in order to deal with the other is not a viable, long-term option.

The first step to achieving an enterprise perspective of change is to have visibility of the different initiatives being considered across the organization. Taking an open, holistic approach ensures the most appropriate combination of changes to be selected for execution and eventual release. Governance frameworks such as Val IT [1] recognize that a portfolio approach is needed to manage the investments that an organization makes. However, conventional attempts at change governance tend to be siloed and undertaken only by specific disciplines (like IT) or in isolated pockets across the enterprise. It is an interesting question: Can the outcome be "enterprise" if the implementation is siloed?

Change governance is best applied at the highest levels in the enterprise and at the earliest point in considering the changes. The principle that guides enterprise change governance is that the clarity of enterprise strategy and the efficiency of its execution starts with the top management. When the change initiatives to be undertaken are endorsed by business executives and senior management, then there is less likelihood of confusion and conflict when these changes are implemented. However, without an appreciation of the factors that constrain an organization's ability to deliver and adapt to change, the stage is set for conflict at the lower levels of the organization.

Bottom-up attempts at change control and governance can appear heavy-handed, bureaucratic processes with onerous approval gates apparently designed to delay progress. These can be symptoms of a misaligned organization. They present the risk that different factions have their own agendas and use governance to achieve local goals over the interests of the enterprise. In such an environment, any reviews and impact assessments have the potential to become time-consuming, document-driven exercises that frustrate stakeholders and result in conflicts and escalations.

There is a false, but widely held perception that governance processes slow the delivery of new business capabilities. Yet, as evident in the case of software, with suitable techniques and tools a release-centric approach can be effective and facilitate greater enterprise agility. Adjustments to portfolio management processes can bring this delivery focus to the enterprise and establish a more open, less siloed approach to how change initiatives are identified, analyzed, and selected for execution. Although strategy plays a crucial role in the selection process, the continuity and stability of the enterprise system necessitate other considerations being brought to bear. While the mixing of strategic and operational changes may seem counterintuitive when a strategic change portfolio is the desired result, this apparent contradiction will be explored in this section.

5.1.1 The Enterprise Change Portfolio

A key characteristic of the enterprise is that multiplicity prevails. Rarely can a single program or initiative, no matter how strategic, represent the entirety of the changes taking place across an operating business at a given point in time. In general, there are more likely to be several strategies being pursued at one time by different business units. Further, operational improvement initiatives and essential maintenance upgrades can be expected to be in place without the formal trappings of a project. Any hope of achieving an overarching view of enterprise evolution and of plotting the future state of what are often loosely coupled departments and business units, requires the perspective to be raised above individual programs of work. A new level of governance is required.

The enterprise change portfolio provides the appropriate level of abstraction to discuss and manage the changes from across the organization. While governance frameworks such as the PMI's Standard for Portfolio Management and the GCO's Portfolio, Programme and Project Offices (P3O) support this contention, they do not require the enterprise perspective to be maintained. Yet the dependencies between different functions in the enterprise can benefit from a consolidated perspective of enterprise initiatives. In this context, the release-centric approach offers a means of ensuring that dependencies between different parts of the enterprise are addressed. An enterprise change portfolio provides the basis for more effective release planning, ensuring that different initiatives across the enterprise are coordinated. Marketing plans can better align with the planned release of new product capabilities and IT schedules can target quiet times in the business calendar. By marshalling enterprise changes and ensuring their execution is carefully planned, governance can itself deliver improved business efficiency.

5.1.1.1 Abolishing Divided Governance

The goal of a more unified enterprise approach to change governance requires us to combine the business focus of traditional portfolio management with the technical rigor of change control. Similarities in governance practices are clearly recognizable across these apparently separate management domains. Generic and consistent processes that define the review, planning, and execution of all manner of changes provides the opportunity to take an integrated view of enterprise evolution. But such a unified approach to change governance necessarily challenges the orthodoxy that attempts to artificially segment and filter the change portfolio in an attempt to focus management attention.

The separate management and handling of change initiatives is commonly accepted practice. Applying the comfortable reductionist "divide and conquer" principle to portfolio managment, frameworks like Val IT recognize that "there are different categories of investments that will be evaluated and managed differently" [2]. However, this conventional wisdom needs to be refined as the enterprise behaves increasing like a complex adaptive system and the project size and budget become imperfect measures of strategic importance or priority. From a systems perspective, eliminating apparently minor, low-priority changes from management consideration too early can constitute a delivery risk. So, while the categorization of changes may be necessary, the governance process needs to be uniform and consistent. Risk is introduced when different governance processes are put in place to manage each category as the siloed handling of initiatives can not only add to administrative overheads, more importantly they reduce management visibility of the changes and their impacts across the enterprise.

The lessons learned in managing software changes apply equally in the devolved yet interrelated systems and processes that serve to glue together the enterprise. Complex systems are intrinsically unstable, and apparently minor and inconsequential changes can have unexpected, disproportionate impacts that can result in potentially catastrophic disruption or failure. The relative ease of changing software and IT systems comes at a price and results in the need for more holistic change governance.

With the appropriate tools, changes can be filtered and sorted to give the different perspectives that currently force different types of changes to be processed differently. However, it is rarely the tools that are deficient but rather their implementation. Indeed, it is common to have several tools to manage the different types of change in the enterprise, so much so that standardizing the change workflow tool can be sufficient justification for a uniform handling of enterprise changes. These change management systems, instigated by different parts of the organization, usually come with their own process, database, and infrastructure, adding to the fragmentation of the enterprise.

In many enterprises today, the governance and oversight of strategic change programs is the province of one or more program management offices (PMOs). Whether serving IT or different parts of the business, they can provide effective local governance over the allocation and tracking of project funds. However, it is increasingly recognized that a consolidated, sometimes virtual body termed the enterprise program management office (EPMO) is necessary to oversee investment spending across the enterprise. While such a body can provide the enterprise perspective that allows funding to be more effectively distributed in line with strategic goals, there is a need to go further and consider nonfinancial dimensions of the change portfolio.

ERM challenges us to go beyond simply identifying and selecting the portfolio and move to tracking of the execution, integration, and delivery of the portfolio. To do so while maintaining system coherence, enterprise releases must take into consideration all change initiatives in the enterprise, irrespective of whether they are strategic and planned or reactive in nature. Aside from the sanctioned programs or projects, there may be a range of activities that essentially modify the enterprise's operating environment and so need to be part of the change portfolio. Perverse as it may seem, delivering priority outcomes in the face of complexity requires us to consider lesser initiatives so that they do not hinder or obstruct business priorities.

To effectively deliver change in the enterprise, the emphasis on change and agility has to be tempered by the recognition that the enterprise is a dynamic system with complex platforms and architectures. Achieving a unified change management approach must recognize the opportunity, and risk, presented by all changes. Release planning across the unified change portfolio is essential to achieving a balanced, optimized enterprise delivery, one that is focused on delivering strategic outcomes and not blinded by the source, scale, or budget associated with the change. Unified portfolio management needs to be democratic in its handling of change and blind to anything other than the best interests of the enterprise.

5.1.1.2 The Democracy of Change

The most common means of controlling change in the enterprise is through the availability of funds. New requests for business change are routinely captured as project proposals, presupposing their execution. When approved by the EPMO, the sponsors progress to developing a business case that aims to quantify the costs, risks, and potential benefits. Portfolio management techniques are then applied to review and prioritize these business cases to determine which projects can be executed within the constraints of resources and budgets. Performing a similar function to that of CCM in the systems engineering domain, portfolio management follows similar processes to those of change control. Yet they can be shown to differ in their core principles, with the key difference being that PPM too often does not recognize the enterprise as a system.

Figure 5.1 presents the different stages that initiatives in a change portfolio undergo. Different terms are often used to describe the different life cycle stages but are immaterial to the current discussion. The number of initiatives are shown schematically as a list being whittled down by successive review gates. The terminology used for these phases can vary depending on the scale of the changes proposed and whether they are candidate programs/projects or enhancement requests undergoing a change approvals cycle. Significant changes

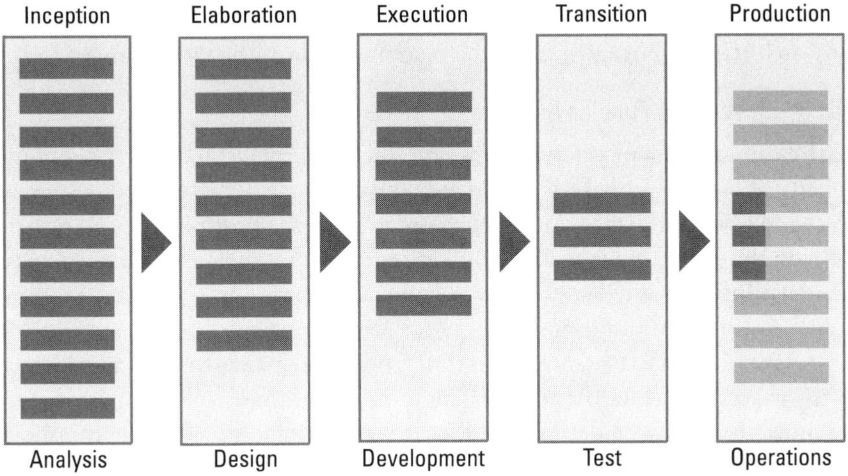

Figure 5.1 Change portfolio.

also have the potential of transforming the business by introducing new systems or replacing existing ones. Irrespective of the scale or language applied, there is a clear correspondence between the activities that have to be undertaken as part of the change governance process. Clearly, all manner of changes have to eventually integrate into the enterprise architecture and become a part of the operating environment. This suggests the principle that, aside from conventions and the somewhat artificial imposition of language, enterprise changes can and should be treated generically irrespective of their scale or nature.

This is not to say that executives need become mired in the review of minor changes—separation of strategic, tactical, and operational changes remains necessary. What is needed is *visibility* of all changes so that those with the specialist knowledge can recognize changes that are related to each other—and raise their execution priority as required. By definition, all changes impact the as-is enterprise architecture in some way. And smaller, apparently insignificant changes have the potential to break the "enterprise system" when they are considered separately from strategic initiatives. Therefore potential impacts need to be carefully analyzed and assessed before they are subjected to traditional management filters. Modern software tools recognized the need to uniformly manage all changes equitably, as shown by the unified change management approach discussed in Section 3.1. There, as in all complex systems, the granularity or scale of the change is not necessarily an indication of its impact.

So, while it is true that management attention is a limited and too-valuable resource to consider all change proposals equally, it is critically important that their attention focuses on the right things. The criteria for categorizing changes therefore needs to be more nuanced and cannot simply be reliant on the traditional blunt measures of budget and duration. It is the number of interdependencies and the potential impacts that deserves equal attention, with early architectural engagement in the planning and prioritization process.

5.1.1.3 Beyond the Funding Model

Portfolio management discipline, the prevailing means of managing the multiplicity of change programs, and indeed the origin of the term itself, is closely related to financial management. Not surprisingly, the portfolio focuses principally on the investments made by the enterprise, and given this financial bias, is often confined to those change programs designated as capital expenditure (CAPEX) for taxation purposes.

As applied in many enterprises today, portfolio management sets the criteria for what constitutes the strategically important initiatives to be included in the portfolio. The very definition of what constitutes a project—the threshold for entry and therefore gaining management attention—is often based principally *upon* the cost of an initiative. If a change has a budget over a certain amount, it will then be considered a project to be treated as a CAPEX initiative subject to the full governance. Any initiative with a budget below the cutoff figure is often, by definition, not deemed a true project. Even where it retains that designation, the change initiative may be categorized as an OPEX initiative and delegated to a less formal governance regime. There may be further budgetary thresholds identified to categorize change initiatives, but the criteria remains principally financial. Subsidiary criteria for filtering or sorting a portfolio can include risk and organizational impact, but these too often end up being surrogates for financial exposure.

Not surprisingly, business cases focus on financial considerations such as program costs and expected returns. The ease of measuring the financial dimensions of an initiative encourages the financial bias over longer term considerations relating to continuity and sustainability. With many businesses reporting financial results quarterly, less tangible aspects of the business such as enterprise coherence and stability can be less evident in strategic decision making. Yet, these more intangible values need to be considered as part of a more comprehensive assessment of the portfolio. A focus on business value can help reduce the emphasis on financial considerations and the value measuring methodology (VMM) [5] makes that point with its introductory document entitled, *It's Not Just Return on Investment Anymore.*

The allure of simplistic assessments of the portfolio (potentially even those seeking to account for business value) is that they enable numeric calculations that allow strategic initiatives to be sifted and sorted like a pack of cards. Initiatives can be considered independently of each other, separately estimated and evaluated, and recombined at will, in different configurations. Those initiatives that have their business case approved or deemed a strategic priority can be separated from the pack for execution. The expectation is that these programs can be independently managed with the assigned program manager responsible for delivering the business benefits documented in the business case—unconnected to any of the other initiatives in the portfolio.

Figure 5.2 shows the common governance gates through which an initiative might go under conventional portfolio management practices. The review process focuses principally on budgets, and filters the change portfolio based on the level of sponsorship it receives, funding criteria, and the availability of budgets. These separate considerations can cause governance and oversight to become fragmented with reporting lines based on how initiatives are funded and confined to the specific PMO accountable for those funds. The financial treatment of the budget, whether it is treated as a CAPEX or an operational expenditure (OPEX), can require different management reporting, processes, and standards. Unless it is strategic, an initiative may not be on the EPMO's radar, and even if it is, reporting would be limited to costs and schedules. It is rare that scope or requirements get visibility at the enterprise level, let alone the dependencies between different change programs.

This highlights the difference between the enterprise perspective and a financial or business one. To achieve an enterprise perspective we must take a systems view of changes in the organization. Simply scaling management disciplines to the enterprise level is not sufficient. Even changing the assessment criteria to be more driven by business value does not address interdependencies nor reduce the execution risk they raise. We need to adjust our methods to the new conditions and, like any system, address the emergent properties to be found at this level. Although overused and too often employed inappropriately,

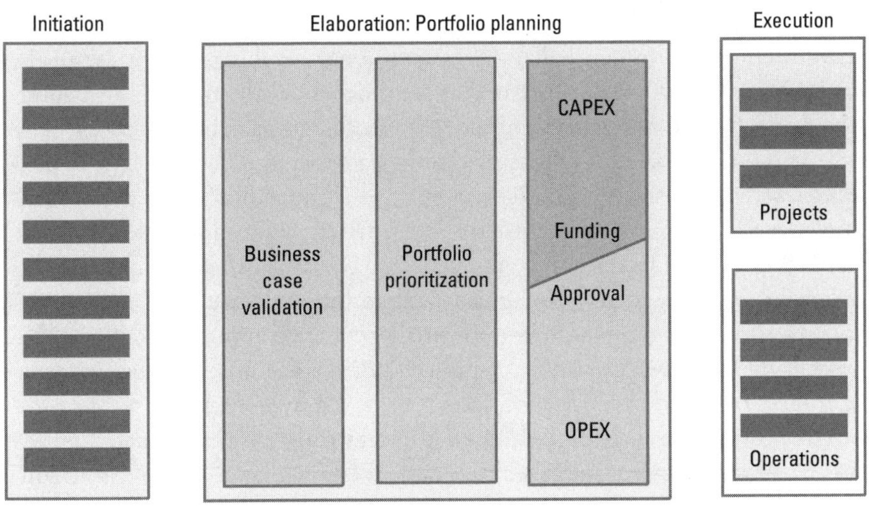

Figure 5.2 Funding approvals process.

Einstein's quote is relevant in this instance: "No problem can be solved from the same level of consciousness that created it."

Instead of considering only the strategic investment options and treating each change separately, there needs to be an appreciation that changes are interconnected with their execution collectively impacting the enterprise architecture. Hence changes cannot simply be treated in isolation and the selection of the change portfolio to be executed must instead comprise a compatible set that maintains the enterprise's coherence. That is to say that changes, and the projects established to deliver them, need to fit into an overarching release framework. To achieve that, a unified approach to managing changes is essential, one that processes all potential changes to the enterprise in a consistent, but not necessarily identical manner.

Achieving visibility of all prospective changes is essential for strategic oversight and governance as it enables synergies to emerge and percolate between what might otherwise be stratified layers of management. From the multiplicity of proposals and alternative future directions for the enterprise, a more considered selection and holistic planning process—one not simply focused on funding— would provide the opportunity to assess the affinity between different changes. Release planning, with this deeper understanding of change allows for opportunistic, optimal combinations of changes to emerge that can be more cost-effective to execute. To achieve this new level of maturity needs portfolio management to be extended to adopt more of a systems thinking approach, considering the enterprise more as a delicate system configuration that needs careful management attention.

5.2 Planning the Portfolio Release

Release-centric planning challenges conventional thinking, which treats the portfolio as a collection of projects that are independently managed. An enterprise portfolio is not simply a collective noun for projects but an entity in its own right—the subject of release planning, tracking, and delivery. The dependencies between changes and project proposals. along with the logistics of planning and executing the release itself is what drives the need for release-centric portfolio delivery. The difficulty of managing a continual stream of changes sees the need to package them into regular planned future release.

The prevailing view of portfolio management commonly confines it to the prioritization and initiation of projects. After that point project management takes over and individual projects are tracked independently. While project level governance may reassess these project's progress at nominated review gates and assign further budget as necessary, the portfolio as a whole is no longer considered. There is little overall coordination of project dependencies like those in the enterprise pattern, nor consideration of the overall impact projects have on the enterprise. By contrast, a release-centric view aims to identify and mitigate risk by reviewing and understanding the system impacts and addressing them in portfolio execution. Instead of independent project releases, the change portfolio is delivered in several coherent tranches or baselined iterations of the enterprise architectures. The inevitability of packaging or clumping of changes initiatives for delivery purposes needs to be recognized early in portfolio planning. If increasing interdependencies between projects force alignment of their schedules and synchronization of releases, then the execution portfolio needs to be designed from the outset to include related changes that can be delivered together in a future release timebox.

The proposed open approach to portfolio management therefore takes a more democratic enterprise perspective, considering a wider range of change initiatives and the end-to-end portfolio life cycle. In this approach, release planning and tracking now must bring together and integrate the different, often interdependent changes. While this may appear to be an overhead that delays the start of approved programs, it is a form of front-end loading designed to avoid future dependency problems that can cause delays and quality issues for the projects.

The early consideration of the release in selecting, planning, and designing a portfolio for execution is the hallmark of ERM and its release-centric perspective. With the end in mind, the future release can be communicated and forms the delivery contract and commits the different supplier teams to a release timeframe or in development language a timebox. Such a future milestone provides some certainty in a volatile business environment and allows all parties to plan their resourcing around these milestones. The collection of future release timeboxes represents the release roadmap that can be shared, as far as possible, across all initiatives in the enterprise irrespective of their scope, granularity, and sponsor. Organizing delivery around grouped portfolio releases rather than individual projects opens the possibility of a more democratic approach to planning releases. Small changes can also target the published enterprise release milestones, which allows them to share integration, test, and transitional change activities with the larger initiatives.

Planning these enterprise release can address the logistics of portfolio execution and ensure a more orderly change transition for the users and customers impacted. The review and selection of a compatible set of change initiatives should be an equally important consideration in the portfolio management, balancing the financial perspective with an appreciation of each release's change impacts and dependencies that can mitigate both execution and user-acceptance risk.

Understanding and addressing the release logistics early in the planning cycle serves to better guide portfolio execution. Targeting future release timeboxes establishes the roadmap for enterprise evolution, providing the opportunity for those impacted to better prepare for the imminent changes. Organizational change management can help ensure business readiness while internal to the release, planning must address dependencies to avoid destructive interference between separate changes that can diminish the promised capabilities.

5.2.1 Enterprise Release Logistics

Selecting the change initiatives that best meet the strategic needs of the enterprise should only be the first step in effective portfolio management. The step that is less frequently undertaken, and often ignored completely, is to address the logistics of delivering the new business and transitioning the enterprise from its business as usual state.

The reality is that releases are expensive constructs. This fact is easily overlooked in the enthusiasm to improve business performance. However, integrating and absorbing multiple changes into the enterprise's operating model takes significant administrative, technical, and management effort.

Here we present the argument that the enterprise release is indeed an inescapable result of increasing enterprise complexity and critical for managing strategic execution and delivery. Treating the release as a delivery vehicle for a portfolio of changes helps reduce overheads associated with frequent and uncoordinated project releases and recognizes the system properties and interdependencies that exist between them. A release-centric approach to planning accommodates the trend towards coordinated project releases, by proactively designing and planning for delivery of a change portfolio as regular, timeboxed enterprise releases. Planners must have cross-disciplinary skills and architectural input is essential to ensure the release timeboxes are optimally populated and deliver successive iterations of the evolving organization via what can be termed an enterprise release train.

5.2.1.1 Release Planning Framework

The portfolio planning and governance effort undertaken can often be disconnected from the execution. Where each proposal or business case presents its own separate delivery schedule it makes it difficult for decision makers to understand the context within which the resulting project might execute. Approving funding for such plans can create an execution quagmire as multiple projects, all approved and endorsed, compete for resources and priority. Clearly, it is not a desirable situation when the time and effort taken by executives making difficult trade-off decisions essentially wasted because they have little visibility of the execution reality

An overarching release planning framework with regular and agreed release timeboxes at least presents shared targets that separate projects can identify, allowing for example the quick identification of periods where there may simply be too many disparate changes for the organization to absorb. Similarly, as many projects in the enterprise rely on shared resources, the delivery capacity of the different teams can be assessed and implicit dependencies identified earlier. Without taking a pragmatic delivery focus, the best strategic plans can only be aspirational fictions that are detached from the reality on the ground. Yet this is a regular occurrence in organizations that detach strategy making and planning from the practicalities of its execution.

Too often, once a business case gets the green light and budget is allocated, all bets are off. The approved project often has the freedom to review everything from the original scope to the design constraints that gave rise to the original estimates. This is not simply because of the lack of governance and while clearly inefficient, such behaviors can be necessary. This period of reinvention of the project is often a result of key individuals responsible for the business case having moved onto different initiatives. Exacerbated by delays in making the decision to proceed with a proposal, those involved in the early inception phase of a project are replaced by fresh personnel. Alternatively, key personnel may only be assigned to develop the business case and are rotated off the assignment and allow a different project team, often with different ideas about the solution, to take on execution responsibilities.

There can be legitimate reasons why the original project planning needs to be reappraised, even if this reinvention can never be considered good practice. For one thing, the original proposal may be superficial and considerations not covered adequately in the early analysis now have to be addressed. But the condition most likely to require replanning and which a release-centric approach helps address, is that other proposals and in-flight projects can invalidate key assumptions made. What is most dangerous in volatile businesses is that the current state of the enterprise is an unknowable, moving target. Such reactive environments only allow change impact to be determine at the point of impact: rendering all planning redundant.

These are all symptoms of treating projects, their business cases and eventual execution as independent entities, instead of being a part of an interdependent portfolio. In essence, independent project planning only works in environments where the enterprise pattern discussed in Chapter 3 is not applicable and that there are few dependencies or impacts on shared assets or resources.

In the multiproject enterprise with several concurrent streams of change, the situation looks something like that shown in Figure 5.3. Irrespective of when a project was initiated, at any point in time several projects will be working to their own schedules and delivery milestones, shown as horizontal streams in the diagram.

While executing such a portfolio of projects may be appropriate when they impact distinct groups or independent businesses (as might be the case for distinct implementations of an ERP system at different client sites), it is

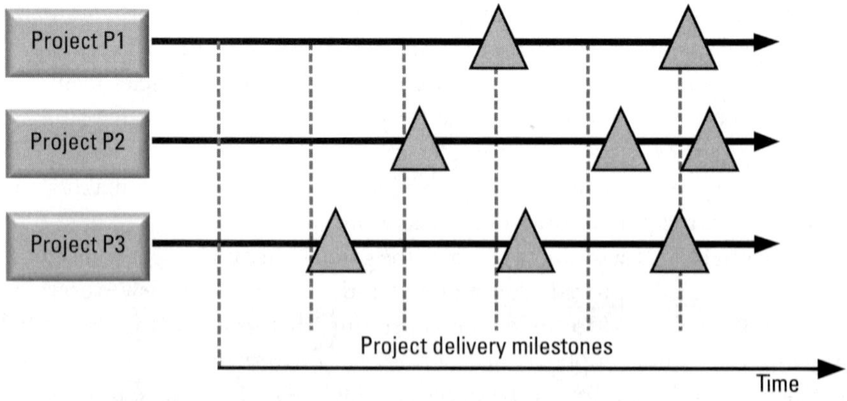

Figure 5.3 Independent project plans.

unlikely to be a successful strategy when their impact is on a single enterprise or on a virtual enterprise with interdependent supply chains that form a market ecosystem. The cumulative impacts of different projects shown in the figure result in continual change that destabilizes the enterprise and quickly becomes unsustainable.

Projects adopting iterative delivery cycles, while reducing their delivery risk, can be seen to potentially compound the confusion by imposing incessant and potentially uncoordinated changes on their customers. Agility applied at the wrong level of abstraction serves only to slow and disrupt the enterprise. This is the enterprise equivalent of the common "throw it over the fence" scenario in development where one team would simply hand over an artifact to those responsible for the next phase whether they were ready or not. Instead, project delivery needs to be sensitive to the people, process, and technology dimensions of the enterprise architecture that must absorb the changes. Inability to manage the onslaught of changes can give rise to uncertainty and inefficiency in the enterprise that directly impacts business performance.

Planning projects within an overarching release framework that strategically guides enterprise evolution can be stablizing. The ability to anticipate future changes reduces uncertainty, allowing business to prepare for change instead of operating in a continual state of flux. Predicting the demand for key resources and spikes in the workload can reduce the need for projects in the portfolio to compete with each other. There are increased demands on governance at the release level that may need to potentially rein in projects and reassess their schedules so they better fit into the strategic direction. This ultimately makes projects and their managers subordinate to those overseeing the enterprise releases. Putting it another way, working within a release framework can reassert the need for organizational change, transitional stability, and business continuity over change imperatives. It offers leaders to fully assess their strategic change portfolio and provides the means to better balance business change and enterprise complexity.

5.2.1.2 The Release Price Tag

The disruption caused by multiple projects separately delivering their capabilities to stakeholders in the enterprise is bad enough, but it is unconscionable when one considers the inefficiency and costs of uncoordinated releases. Aside from the impacts on the organization, the planning, integrating, testing, distributing, and communicating associated with a release often require significant administrative and management effort.

Put simply, managing projects and their iterations independently increases the cost of change. While the benefits of iterative delivery cycles are clear and well argued by the Agilists, there are clearly logistical costs incurred. Aside from those stressful internal coordination and integration activities, there are the external impacts to the customer or end users. Implementing upgrade programs resulting from even relatively minor releases can require communication of new or changed capabilities, associated process adjustments, training for users, and potential modification to interfacing systems. Releases are therefore a costly exercise that incurs administrative, technical, and management overheads. Even where changes are simple to implement, as in software development, the consequential impacts of these changes can be anything but simple.

The ability for an enterprise to adapt to and absorb change is less than its ability to generate it. This reality is often overlooked by the ardent supporters of agile methods who take only a solution-centric view of development. While automation tools make the continuous integration of software the panacea for the challenges of parallel development, there is no easy solution for parallel projects in the enterprise. The ability to integrate different changes, identify any resulting problems, refactor a poor design, and redistribute the fix represent some of the unique features of software. While a significant challenge in its own right, these options are simply not available for change in the enterprise. Here the impacts to the people, processes, and technology dimensions of the enterprise architecture are simply not that forgiving.

The stakes are much higher when considering changes in the enterprise. The logistical costs of releasing new capabilities are more akin to that of a physical product that has to be manufactured, packaged, distributed, displayed, and marketed. The supply chain to arrive at the desired solution and the relationships and lead times associated with each step in the manufacture and distribution process should be analyzed and optimized, as they are in lean manufacturing. Yet in the enterprise it is common for strategic execution to be managed simplistically, monitoring only the measurable parameters of cost and schedule of individual projects.

To achieve an agile enterprise, as opposed to simply agile projects, requires deeper insight and more innovative techniques to provide an integrating perspective of multiple, concurrent changes. Enterprise releases must address the need to transition individual users, stakeholders, teams, and sometimes the entire organization as part of transformative programs. Performing different activities in the enterprise, each group has its own resistance factors, mobilization times, and change latency. Their processes have their own unique time scales or can follow a regular business cycle, for example monthly, quarterly, or financial year reporting. Some changes are only relevant or visible once a cycle and capability become useful at that point. For example, an enhancement to a customer bill is only evident in the monthly billing cycle. The delivery of any new system capability therefore needs to be attuned to this temporal reality.

Figure 5.4 shows the implicit dependencies between projects and how they cannot be relied upon when projects have uncoordinated project schedules. The current architecture at the time the middle project (P2) is planned, designed, and estimated can be seen to be different by the time it comes to its first delivery milestone. The other two projects (P1 and P3) that have their releases before it, can inadvertently cause the assumptions made at the initial baseline to break with unexpected results discovered as P2's own release approaches.

The resulting uncertainty means that quality problems emerge at execution time, possibly for no fault other than the fact that the original design decisions have been invalidated by other projects impacting the same components. The lack of confidence in planning decision making in such a volatile environment results in reactive behavior where nothing can be trusted. The lack of reliable foundations wastes time and effort, but even more important, erodes trust in any management process.

This again is reminiscent of the software problem without effective management processes like SCM in place. Unstable or unreliable baselines can require brute-force testing as uncontrolled changes break previously stable features or overwrite previous fixes causing earlier problems to reappear. Projects

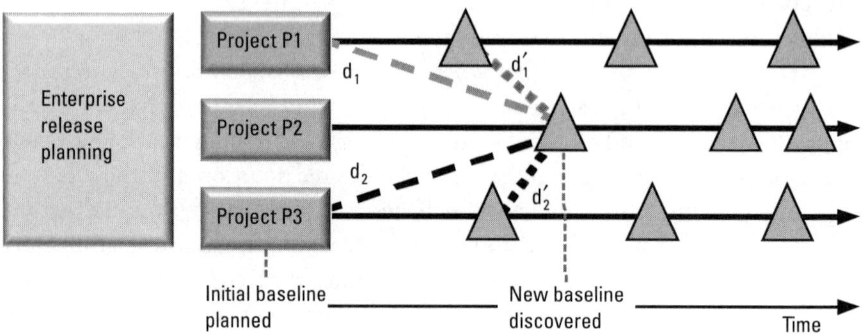

Figure 5.4 Changing project baselines.

in the enterprise can suffer from similar churn and inefficiency without the benefit of software's CM, build, and integration tools—maintaining a single source of truth or a baseline of the entire enterprise is not easy even if it is clearly an aspiration of enterprise architecture. The chaos that can result from continual, uncoordinated changes needs to be tempered by a paced enterprise release cycle with more manageable impacts. To be effective in the execution and delivery of a change portfolio requires equivalent toolsets supported by new techniques to improve visibility of the changes and manage the fluid baselines of the enterprise.

5.2.1.3 Enterprise Release Roadmap

A simple remedy for enterprise churn is to have fewer changes and indeed shutting off change is a common, if somewhat blunt, governance mechanism. Stop everything, or at least slow it down until "order" can be restored, was once the stated goal of many CM shops. And while it has the benefit of reducing both churn and the costs and disruption associated with change, it may not do much for the competitiveness of the business.

A more nuanced approach is to develop an enterprise release roadmap that attempts to consolidate the separate project releases into fewer, integrated enterprise releases. This helps to reduce the disruption of separate projects impacting the enterprise in an uncoordinated fashion and, with the ability to address dependencies with an integrated release, improve reliability and quality of the outcome. Further, consolidating the number of project milestones can dramatically reduce the costs and logistics associated with integration and testing releases.

The key principle which emerges is that these enterprise releases transcend individual project schedules. Projects now have less flexibility to set their own delivery schedule and need to justify deliveries outside the shared enterprise release calendar. Much as a larger program can set the schedule for smaller projects within its ambit, the enterprise release roadmap establishes key milestone that projects need to work to and synchronize their schedules with.

Beyond simply consolidating and integrating several projects into the enterprise release, there is a trend towards a regular cycle for releases. The benefits of a predictable release cycle offer predictability and respite for those charged with delivering a continual string of releases as well as those receiving them. Depending on the nature and volatility of the business the release cycle can vary, but the regularity allows both the business and the separate delivery teams to better plan and prepare for each release.

Software development projects often use techniques like continuous integration to provide complete builds of a product on a weekly, and sometimes even a daily basis. While only some of these builds will be selected to undergo the formality and ceremony of an official product release, the mechanics are in place to allow a complete, new version of the product at a regular and reliable pulse. In the enterprise, the ability of the organization to absorb and adapt to change makes even a monthly release cycle aggressive with a quarterly, biannual, or even annual schedule of regular releases more likely.

Having a regular release schedule is closely related to the timeboxing technique that is the basis of agile methods like Scrum [3] but which goes further back to early IT software methodologies like rapid application development (RAD) [4]. This technique ensured that completed software changes would not be moved into production as they completed and instead had to be scheduled for delivery in a timebox. The imposition of regular releases places an additional constraint on standard iterative development and has the benefit of providing predictability. This comes at the risk of having some resources, such as testers being sidelined as they await their role to be called upon, something which is easily remedied by having multiple releases at different stages of their life cycle being worked on concurrently.

Applied to projects instead of smaller granularity changes, timeboxing does not require a major leap of faith. The necessary abstraction simply replaces the release of the software's architecture with that of the entire enterprise architecture. The result is that changes to the enterprise can occur in more controlled tranches, or pulses, that can be matched to the enterprise's ability or willingness to accommodate change. The resulting releases can be selected and designed to accommodate system-like behaviors that result in a more cohesive and complete whole of business solution.

Switching to aligned or synchronized enterprise releases is not a move that can be taken lightly. Project sponsor and stakeholders accustomed to setting their own delivery schedules have to accept the compromise of working within a timeboxed release framework. As Figure 5.5 shows, instead of projects simply working to their own independent schedules, they now have to fit into the release roadmap and target the nearest available timebox. Some projects (such as P1 and P3 shown in the figure) need only make minor adjustments to their delivery schedule while some (like P2 in the diagram) may require renegotiation with stakeholders to shift delivery milestones and reset expectations.

Of course it is not essential for all projects to fall into line with the enterprise release calendar. While there are clear efficiency and change transition benefits that come from aligning with a release timebox, there can be good reasons why a project may choose to remain outside of the enterprise release calendar. Synchronizing project releases with regular enterprise release windows is only essential when the impacts are too common or shared components of the enterprise architecture. There is less benefit where a business unit operates independently of any shared platforms or a project has no reliance or dependency on other initiatives or common resources. Deliveries outside of the mainstream enterprise framework need to be justifiable and with projects increasingly

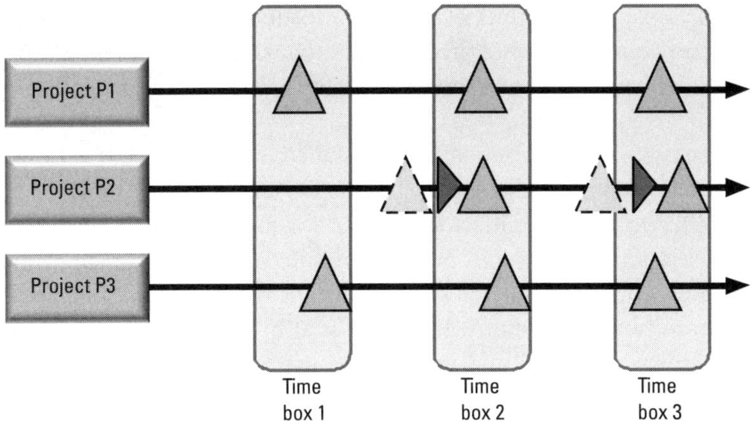

Figure 5.5 Enterprise release windows.

operating within a network of implicit and often vaguely understood dependencies, there is less risk in simply adhering to a shared release calendar. Where changes can safely have their own release plans and schedule, they will bear the full costs associated with integration, test, transition, and acceptance of their solution—something that may be sufficient to consider utilizing the release management services offered when adhering to the enterprise release calendar.

Enterprise release planning requires strong governance and senior management support to establish an agreeable enterprise release calendar. Determining the appropriate frequency of the releases and adjusting project delivery schedules to fall in line with them is a compromise not easily accepted by powerful stakeholders. There is likely to be a perception that essential business changes will somehow be delayed and management flexibility lost when a regimen of regular releases is first proposed. Therefore, achieving an enterprise release framework requires recognition of the efficiency and stability benefits at the highest levels in the organization. Like all front-end-loading activities, the challenge is to temper the demand for immediate gratification with confidence that regular timeboxed enterprise releases are more cost-effective, reduce execution risk, and serve to reduce change disruption in the organization. Leaders have to accept that increasing complexity means that any changes proposed to the enterprise need to assessed and regulated as if it were a delicate system. Only then can the enterprise transition smoothly from one release to another—and evolve in accordance with a defined strategy.

5.2.2 Establishing Release Coherence

A release-centric approach to change recognizes that increasingly shared enterprise architectures mean that fewer changes can be managed independently.

Instead, the separate initiatives selected for execution needs to be parceled into coherent and compatible enterprise releases that combine relevant changes irrespective of their granularity and, for the most part, adhere to a calendar of regular, timeboxed releases.

But to achieve release coherence, portfolio governance needs to be extended to include a multidisciplinary design perspective. This is necessary to ensure that the different change initiatives selected for execution are compatible and are not at risk of fracturing the enterprise architecture and increasing complexity. The release can be considered to be an instance of what enterprise architect's term the transition architecture. As the basis for business operations, each stage in the enterprise's architectural evolution must be assessed to be sufficiently stable and reliable—ideally long before the release implements it.

To ensure that the decisions taken at the portfolio level result in the delivery of a new but stable platform for business require the early involvement of enterprise architects—or more specifically architects with an understanding of each of the different dimensions of the enterprise as discussed in Section 1.3. Reviewing proposals and business cases from each of these perspectives allows a better assessment of change impacts and yet this good practice needs to go beyond traditional project thinking to work at the EPMO level and establish a new model of governance and coordination. This would focus not on the individual solution architectures developed for the separate project initiatives but instead look at the sum of those aspects of the solution that are to be delivered in each release timebox. Consolidating the different part solutions expected to be delivered by each project provides the transitional architecture that will in fact be implemented as a part of that release. It is this release architecture that needs to be coherent and stable, making it deserving of stricter architectural governance. Sadly, less attention is usually paid to its definition, review, and assessment than is paid to blueprints representing future-state enterprise or solution architectures that may never actually see the light of day.

5.2.2.1 The Release Architecture

When a PMO focuses on the form of an initiative, considering the critical dimensions of costs and schedule, there is a tendency to treat the projects as black boxes. The solution and capabilities delivered by the project—the actual content of the box—can become secondary as the initiative makes it through the governance process. The leadership assumes that the business case will be delivered according to plan, even as priorities are juggled and costs negotiated with suppliers. Even where the scope and eventual solution is understood and managed, considering each black box in isolation, separate from the other boxes being planned represents a delivery risk.

It is at the point of execution that the contents of the boxes become an unavoidable concern. As was previously shown, the boxes can be expected to have dependencies on each other, bringing separate design decisions into question. It is these design dependencies that break the project paradigm, and unlike the estimates, contingencies, and business benefits that are amenable to simple, arithmetic calculations and can be summed, sorted, and filtered, the disparate solution designs, each in their separate box and approved for execution, are not so easily combined. Sadly, solution architectures are not easily merged.

Several project proposals may be given the green light because their business cases align with strategy (if not always with each other) but since their solution designs are determined independently, there is a risk that they are inconsistent and possibly incompatible with each other. Their potential to negatively interact or interfere with each other or potentially corrupt the enterprise architecture will only be discovered later in the project life cycle—possibly when it is identified by defects.

While each individual business case requires a stand-alone design to allow the estimation of effort and costs, project approval should not be seen as a validation of this high-level solution design. Instead, the separate solution architectures that form the execution portfolio need to be integrated to define the transitional architectures expected to be delivered in each release. As an enterprise evolves and its complexity increases, it is not enough to simply review and approve project (or worse, more granular change) designs in isolation. Instead, a consolidating perspective with strong architecture governance is needed. A design authority for the enterprise needs to be established, with stakeholders representing each of its five dimensions to ensure that an appropriate transition from current state to target state is achieved.

Governance therefore needs to go beyond passively tracking the obvious parameters of the portfolio to become more active and predictive with the planning and design of the future enterprise. Foreseeable obstacles and risks to a release need to be preempted, for example in the organizational dimension where multiple changes can impact users and stakeholders. Even as each project may make preparations for the needed changes to roles and responsibilities, the cumulative effect of multiple changes often results in a flurry of confusion, resulting corrective actions and costly workarounds. The relatively new and burgeoning field of organizational change management, or OCM, is evidence of the need to better understand the impacts of new processes and IT solutions on the people dimension of the enterprise.

The impacts of a portfolio therefore be on the enterprise must of course be considered across all architectural dimensions. Establishing a release architecture, which integrates the separate (aspects of the) solution architectures to be delivered together, is a valuable contribution that enterprise architect's can make to ensure the delivery of strategic change. The role of the enterprise architect in decision making can benefit from this temporal, delivery focus and establish the transitional release architecture in support the enterprise release roadmap. This requires dependencies within the portfolio to be understood and their risk mitigated by planning compatible releases that minimize business disruption. Architecting each release requires careful introduction, as it can result in recommendations to redesign or reschedule certain elements of the portfolio for a subsequent release in order to reduce dependencies. Full, end-to-end, ERM calls for release-centric governance to complement financial governance and ensure a realizable enterprise release roadmap delivers on strategy. Such a perspective addresses real scenarios and designs that directly impact the enterprise, as opposed to the more idealized deliberations of future state enterprise and solution architectures that may never be delivered.

5.2.2.2 Designing the Enterprise Release

Distinctions need to be drawn between the various architectures that become relevant at different points in the change life cycle and levels of the enterprise. The distinction between the enterprise architecture, solution architectures relating to proposed changes, and the system architectures for each of the solution components are broadly understood, if not always applied. While the matrix notation to be introduced in Chapter 7 goes some way to differentiating these different architectures, rarely do we encounter the release architecture being discussed in an organization. Even though it is essentially what enterprise architects aim to define in their target-state blueprints. With all different architectures to develop and maintain, it is no surprise that little attempt is made to capture the interim, temporal states at point of delivery. As essential as such an architecture can be the stability of the enterprise, architects are usually scarce and expensive resources who are directed towards more important duties. Usually this entails evaluating exciting new change initiatives rather than considering the relatively dry reality of whether a random set of initiatives underway at any given time can deliver a compatible and stable business platform. If the enterprise can be considered a system, then this oversight is incredible. We have essentially let the enterprise evolve with little real governance applied at the level it matters.

The introduction, or more appropriately the elevation, of the release architecture can ensure that the operational configuration of the enterprise as a whole is managed. This is configuration management of the enterprise, and while it may be a lot less exciting than designing new solutions and executing projects, it is the essential discipline that we have allowed to become fragmented. As a configuration, it does not suffice to simply manage parts of it, it is the interdependencies that create the rich and unpredictable behaviors that make the difference between success and failure. And as in the system engineering context, where best to protect that configuration from incoherence than at the point of change.

Figure 5.6 shows the point at which release governance should be inserted into the portfolio planning process. Even as project proposals have their

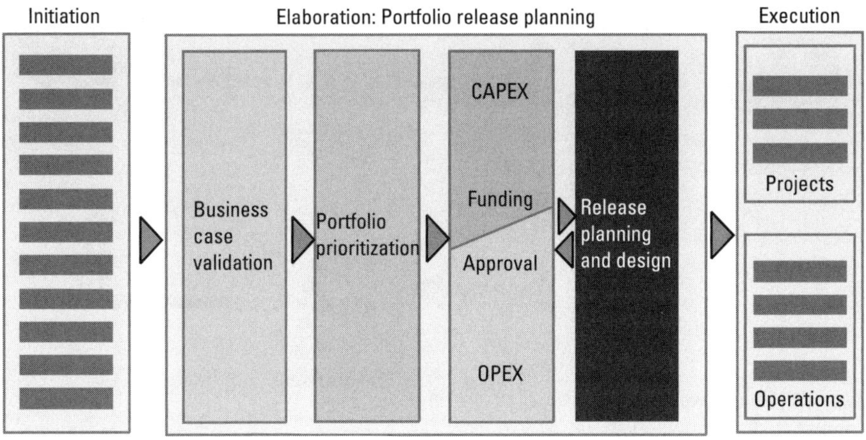

Figure 5.6 Enterprise release planning .

business cases approved and funding made available, the actual delivery schedule needs to be vetted as part of release planning and design. Acting as a change control board would in governing changes to a complex system, an enterprise-level strategic delivery body should determine the design implications from an integrated portfolio perspective. This is the body that should be the custodian of the enterprise release roadmap and as a result can best gauge the impacts on the organization resulting from the approved change portfolio. Where needed, initial funding may be suspended, with individual projects needing to redesign or modify their delivery plans before being given a release window.

As such release planning and design should not simply be an information gathering activity that rubber stamps projects—although even that could be of tremendous benefit in providing visibility of upcoming changes to the enterprise. The goal is not simply to package already architected and planned solutions into a release. Instead, there should be a two–way dialog as shown in Figure 5.7 between projects and those managing the release schedules. Rather than being seen as a constraint on projects, effectively conducted release planning and design can facilitate the delivery of individual solutions, and even offer the potential of cost reductions through shared and consolidated designs.

With all the different architectures to develop and maintain, it is no surprise that little attempt is made to capture the interim, temporal states at point of delivery. As essential as such an architecture can be the stability of the enterprise, architects are usually scarce and expensive resources who are directed towards more important duties. Usually this entails evaluating exciting new change initiatives rather than considering the relatively dry reality of whether a random set of initiatives underway at any given time can deliver a compatible and stable business platform. If the enterprise can be considered a system, then

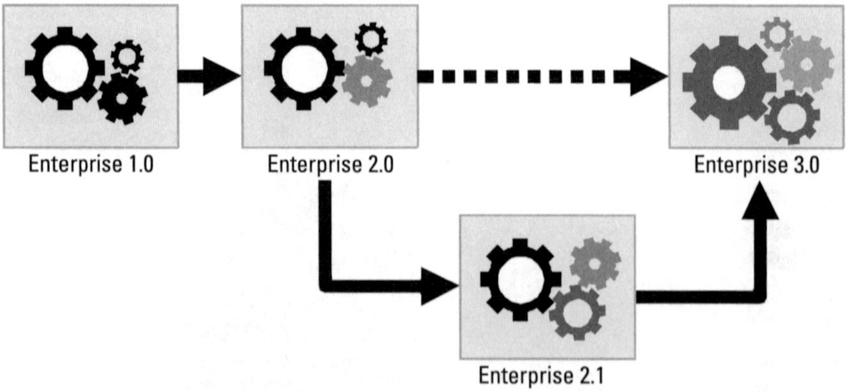

Figure 5.7 Enterprise release roadmap.

this oversight is incredible. We have essentially let the enterprise evolve with little real governance applied at the level it matters.

Having a release architecture presupposes a knowledge of the projects, and their solutions, which will form a part of the release. A reliable enterprise release roadmap needs to be in place—similar to that presented at the beginning of this book in Figure 1.1 but including scheduled release dates—so that projects can align their schedules with the enterprise release calendar. Each project needs to target delivery of capabilities to specific releases so the tranche of concurrently executed projects can synchronize their separate delivery schedules to participate in shared integration, test and release activities. In this way, the identification of the interdependencies can be addressed at the outset and not simply be an afterthought, often left to be "discovered" late in the project or release life cycle, possibly as defects. Maintaining such a visible release roadmap is a key service which ERM can offer projects for their participation in enterprise releases. It can help to prepare stakeholders and end-users for change by offering consolidated communications, as well as ensuring that integration and test resources are available and mobilized in readiness for the projects' deliveries.

The results of release planning and design are shown in Figure 5.7, as a fleshed-out roadmap that has a defined release architecture for each future release. Achieving such a roadmap requires close cooperation with projects to achieve a sufficient level of detail to mitigate the risk of inconsistencies or incompatibilities across project deliveries. The planning requires a multidisciplinary review of the separate solutions that can involve business strategists, architects, and planners. Such a discipline applied to a change portfolio, and working in conjunction with the EPMO's funding approvals cycle, can establish an enterprise release roadmap that guides the strategic evolution of the enterprise from its current state to the desired future state. With such a governance mechanism, investment choices can be made with due regard to execution feasibility and the risk to business disruption.

Establishing a release architecture ensures that there are fewer surprises and that its preparation need not be onerous. The different solution architectures proposed by projects targeting a specific release need to be reviewed collectively. This happens naturally, albeit informally, when the architects are a shared resource working across multiple projects. However, when the traditional solution architecture review takes a wider, release-centric perspective it has the potential to optimize design and save costs. Since project dependencies can affect design decisions, it can be wasteful to develop the detailed designs independently. Instead, the release planning step should take the preliminary designs as input and proceed to review them, merging them into an integrated, coherent release architecture. Where design incompatibilities are detected project schedules may need to be replanned and estimates provided as part of the business cases can be expected to be revised as the design is altered.

The intimate relationship between the release design and execution necessarily make the release planning activity a shared responsibility of business and technical management. Developing a release architecture needs to be an integral aspect of the planning step identified in Figure 5.6. It prepares the groundwork for the successful execution of the different initiatives and supplements the business prioritization and financial planning activities conducted by the PMO. The enterprise release provides a feasibility or risk analysis of the combined changes proposals and, in Chapter 7, we will describe how it can be used to optimize resource utilization and increase the throughput and productivity. With the appropriate techniques a release-centric review of the initiatives introduces the possibility of reducing estimates by identifying opportunities for design efficiencies and reuse across the portfolio.

Release-centric considerations would be familiar to those with systems engineering backgrounds, but such thinking has not been widely accepted in business and IT. Instead, the convenient and optimistic simplification has been adopted that views each business initiative being managed in isolation, with the planning, design, and execution conducted independent of concurrent change activities. This fallacy cannot be supported in the complex, integrated enterprise. Instead, simple logic demands that a correspondingly integrated management framework such as ERM be put in place. The need for integrated systems thinking transcends both IT and business, making irrelevant the usual distinctions and sometimes conflicting perspectives of each camp.

References

[1] Shpilberg, D., et al., "Avoiding the Alignment Trap in IT," *MIT Sloan Management Review,* Vol. 49, Fall 2007.

[2] COBIT 4.1. *Control Objectives for Information and Related Technologies,* IT Governance Institute, 2008.

[3] Schwaber, K., *Agile Project Management with Scrum,* Seattle, WA: Microsoft Press, 2004.

[4] Martin, J., *Rapid Application Development,* New York: Macmillan, 1991.

[5] The Value Measuring Methodology (VMM) - How-To-Guide, Federal Chief Information Officers Council, Best Practices Committee, http://www.cio.gov/documents/TheValue-of_IT_Investments.pdf, 2003 (retrieved Aug 2011)

6

Delivering the Enterprise Release

A man has free choice to the extent that he is rational.
—Thomas Aquinas, philosopher and theologian

In the preceding chapter we looked at how the strategic planning should take place, with a top-down governance model that attempts to control the investment budgets that drive strategic change in the enterprise. To adopt a release-centric approach meant going beyond strategic planning to packaging together initiatives as part of release planning. A high-level release architecture is needed to ensure feasibility and temporal coherence of each enterprise release that delivers another iteration of the strategic vision, the latest version of the enterprise.

Strategic planning and portfolio governance must take the execution landscape into account. Simply approving initiatives without knowing the environment in which they have to be executed forces concurrent projects into competitive interference that introduces execution risk and delays outcomes. Even when projects are initiated after careful portfolio prioritization and difficult trade-offs, the capacity of the organization to deliver the benefits in the business case is severely compromised if the change portfolio is incomplete or does not consider concurrently executing projects that can confuse baselines and invalidate key assumptions.

As illustrated in Figure 6.1, the link between strategic planning and execution is a vital, yet often a weak link in the fast-paced, modern enterprise. While portfolio governance is viewed as the preserve of the board and senior leadership who appraise strategic initiatives early at the inception stage, the latter, execution and delivery stages of (both past and present) approved projects is delegated to lower levels of management. The disconnect between the two phases of an initiative and the tiers of management who are responsible

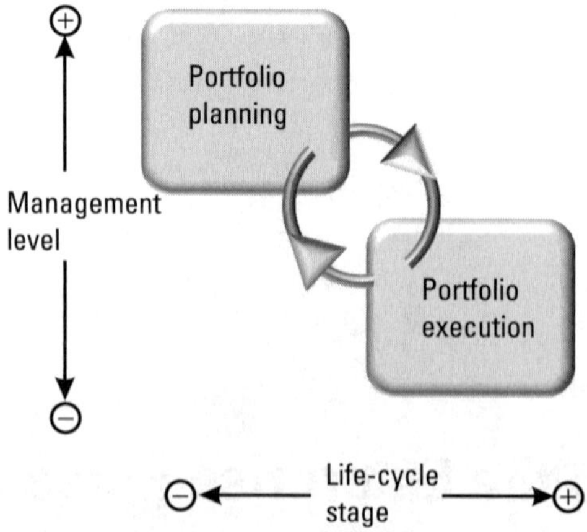

Figure 6.1 Connecting planning and execution.

for discharging it is a key contributor to the disappointing success rates in the enterprise. The point where the strategic intent of the leadership team crashes into the stark reality of execution gives rise to the chaos that can characterize projects in the enterprise.

When connected, the top-down business planning perspective and the bottom-up capability delivery perspective can result in a clash of philosophies that give rise to the conflicts and tensions that characterize the business-IT relationship, as just one example of the divide between intent and execution. Business alignment has been the mantra for decades as competitive market conditions have demanded continual change with the organization expected to be driven by that imperative. However, increasing consolidation of the enterprise architecture driven by competing, cost-reduction strategies have been shown to make delivering to business expectations a growing challenge. This impasse has been the basis of a major study of IT efficacy [1] that identified what the authors term the alignment trap, where a singular focus on strategic alignment was shown to correlate with greater IT spending and lower growth companies—especially when they were ineffective at delivering projects on time and on budget.

This chapter therefore sets out to discuss the alignment trap in terms of the reality of working in the enterprise and identify pertinent management issues that contribute to inefficient execution practices. The relationship between customer and supplier is then reviewed and the enterprise pattern deconstructed to explore changes in delivery patterns and their corresponding management approaches. The release-centric approach to planning and execution is shown to provide a response to the alignment trap and lays the foundation for an enterprise-scale change competency.

6.1 Strategy and the Alignment Trap

Business alignment describes the extent to which support functions like IT respond to business strategy and organize their resources to address these priorities. The goal of alignment relates to doing the right thing, picking the right strategic initiatives so that time and resources are not expended on activities that provide lesser business value. Alignment, however, relies on the assumption that the organization has the capability to deliver to business expectations—to match doing the right thing with doing the thing right.

According to the popular Warren G. Bennis quote: "Leaders are people who do the right thing; managers are people who do things right." However, it serves us well to understand that neither is sufficient by itself. Further, the distinction between leadership and management, when combined with personal ambitions, results in most individuals aspiring to leadership and neglecting management. This can result in a gulf between the strategic plan and the organization's ability to execute and deliver it.

The rapid rate of change may have a role to play in disrupting the balance between planning and execution, but it is somewhat ironic that while execution needs to be aligned with strategy, strategy is too often detached from the realities of its execution. Yet strategy cannot just be about unrestrained desires; to be realistic it must consider the means of executing and delivering that strategy. Ultimately, successful delivery of strategic outcomes relies on having the capability to execute.

6.1.1 Opportunities and Traps

The alignment of business and its supporting functions has a long tradition, particularly in IT strategy and governance. IT management, attempting to control its budget that covers both the maintenance of existing and the procurement of new software solutions, relies on alignment with strategy as an important tool to ensure business planning is tightly coupled with with enabling IT capabilities and services.

It is in the support of this cause that the stream of changes demanded by business have seen the ascendance of the project paradigm. It has also had a subtle and disturbing trend where the business is ultimately the customer and the supporting services must somehow find a way to meet its demands. This conveniently simplistic viewpoint can result in the denial of complexity and a can-do attitude that expects all things to be possible with sufficient motivation, dedication and innovation. While stirring, this can represents a return to the cowboy era of management that focuses on short-term capabilities over more sustainable solutions.

The success of agile methods illustrates in large part the demand that exists for addressing the business need while changing the rules as to what it means to "do it right." Agility is a bottom-up response that attempts to short-circuit the management chain so that strategy can be communicated more directly to those who perform the desired function, in this instance software development. In order to meet the demand for rapid delivery of new capability, agile methods literally needed to twist key management conventions from traditional planning and top-down design, architected solutions to more incremental and adaptive solution designs. This required the reengineering of key development practices and juggled the order of unavoidable constraints such as scope, costs, and schedule. Architecture gave way to a more emergent design philosophy, with techniques like refactoring possible only because of the peculiar features of the software medium.

If necessity is the mother of invention, then the emergence of agile methods might have been predicted in software development. However, there is little capacity in the enterprise to absorb or adapt to the onslaught of changes. The opportunity to improve agility in the enterprise will require different, more holistic techniques to address the combination of both change and complexity found in the fourth quadrant (see Figure 2.4).

Agile techniques such as refactoring are not available in the wider enterprise and continuous releases are disruptive and not economically feasible, even if it were possible. Flexibility and adaptability in the enterprise resides in management; the processes used to govern, plan, and execute change. This is what constitutes the software that runs the enterprise and where the opportunities, and traps, for improving the delivery of strategic outcomes lie. It is these processes that have to be tuned to redress the balance between strategic planning and successful execution, with the enterprise release being the focus of communication and negotiation.

6.1.1.1 Interpreting the Alignment Trap

It is evident that management practices must adapt to address the continual demand for change in the enterprise. Clearly, management professionals have a responsibility to challenge the viability of business strategies and demand the resources and support needed in order to execute them successfully. Even when working in the service of business, experts with specialized knowledge in IT and other support functions have a professional responsibility to consider the consequences of any requirements and to advise business managers of the impacts and risks of expedient decision making.

In today's competitive landscape, however, questioning of strategic initiatives has to be done delicately as it is liable to be unpopular with business sponsors. If an opposing voice is interpreted as criticism of the strategy, or worse, an inability to deliver, the sponsor has the flexibility to shop around till they find individuals or organizations more compliant and willing to do their bidding. In such circumstances an aggressive service provider can always be found who will deliver the immediate requirements without considering the longer-term consequences of a siloed solution on the enterprise architecture. No amount of entreaties, promises, or negotiations can avoid the constraining reality of the shared resources and complex interdependencies found in the enterprise pattern. Unwittingly, in enshrining business alignment we have created a culture that accepts a low success rate and embraces the turmoil and confusion of project execution as if it were a mark of distinction. The need to adopt a more balanced and pragmatic view of enterprise evolution was never more evident.

A study conducted by Bain & Company [1], henceforth referred to as the Bain Study, has shown that having IT focused on implementing business driven strategic initiatives can lead to, what they term the alignment trap. Their conclusion is made clear in the report's subtitle: *Aligning a Poorly Performing IT Organization to the Right Business Objectives Still Won't Get the Objectives Accomplished.* They found that simply attempting to implement the business strategy was not enough. There had to be a matching capability to deliver and execute that vision.

The Bain Study discovered a simple pattern behind the disappointment and failure endemic in IT projects. They found that alignment with business objectives without the corresponding project management capabilities resulted in increased IT spending and poor business growth. Figure 6.2 illustrates the findings of the study, relating alignment and efficacy on two axes and displayed in the customary two-by-two grid. In the *x*-axis is the level of alignment with strategy that is related to doing the right thing, while efficacy is measured along the *y*-axis and is associated with doing the thing right.

While the majority of survey respondents were in the neutral maintenance zone (where "IT projects were treated like plumbing"), 11% was found to be caught in the alignment trap. Even as these companies were found to spend more than the average company on IT, they posted the lowest revenue growth. The remaining two quadrants showed better results with those focusing on running IT projects efficiently (the Well-Oiled IT at bottom right) bringing IT costs down 15% less than average with revenue growing at 11% above average. The quadrant at top right, entitled IT-Enabled Growth, indicates companies that were both effective at managing IT projects and at aligning them to business objectives. This quadrant represented the most successful organizations which showed a 3-year average sales growth exceeding the average by 35% while their costs were 6% lower.

The Bain study captures the business value of maintaining a high level of strategic alignment together with high efficacy in core IT capabilities. This result has a close relationship to the change-complexity analysis in Section 2.2 (aside from the use of the 2x2 grid). The corresponding quadrants between

Figure 2.4 and Figure 6.2 can be seen to represent the management challenge on the one hand and the opportunities offered in overcoming it.

Change and complexity can be considered the fundamental drivers that create the environment in which the enterprise has to operate, the stimulus that begs a response. Seeking alignment can be seen as management's response to change, with a focus on driving through initiatives, based on the current business imperative. Certainly alignment is not as great a challenge if it were to be static over the years allowing alignment to simply be a result of gradual maturation of solutions. Instead, efficacy (in this instance of IT) is at least in part a measure of management's competence in handling increasingly dynamic and complex enterprise architectures.

While low levels of alignment and efficacy require a range of skills and capabilities, advancing to the highest level of each requires an aptitude for managing the multiplicity of interdependent initiatives represented by the previously explored enterprise pattern. With this interpretation, it should come as no surprise that the thinly populated top-right quadrant in Figure 6.2, where the open challenge of managing both change and complexity resides, also offers

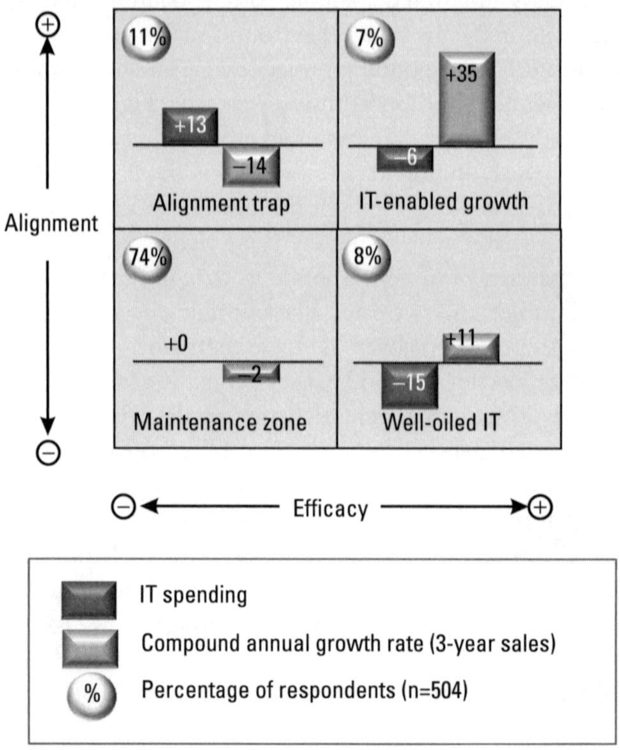

Figure 6.2 Alignment versus efficacy.

the greatest potential for business return when these forces are mastered. This combined challenge conspires to prevent more organizations from attaining the benefits of IT-enabled growth. Even as the rewards can be great, the present state of management practices thwarts the best efforts of many otherwise competent organizations.

6.1.1.2 Qualifying Opportunities

The Bain study is important because it provides current empirical data that informs decision makers on the relative priority of addressing alignment (and, by extension, change) over efficacy in managing increasing architectural complexity. The key management question that the authors address is whether alignment and a focus on strategic planning can be relied upon if there is not a corresponding execution and delivery capability.

While the answer is to some degree common sense, the results of the Bain study clearly point to a management response that favors addressing efficacy over alignment. The analysis showed that efficacy is at least cost-effective, and that [1]:

> For the majority of companies, the single most important task has been to forget about enhancing alignment for the moment and to focus first on increasing the effectiveness of the IT organization. In order for IT to enable growth, that first move is critical.

The study also offers recommendations for avoiding the alignment trap that can be seen to be critical to overcoming the challenges found in the fourth quadrant. The three remedies they propose are to emphasize simplicity, right-source capabilities, and create end-to-end accountabilities. Each of these is pertinent to making the case for ERM.

The alignment trap can be viewed as a consequence of neglecting complexity and a corollary to the change and the complexity cycle discussed in Section 2.2. Alignment is not sustainable and a continual stream of business changes can drive an unwitting enterprise towards increasingly fractured, ineffective enterprise architecture. The authors point to a fundamental misconception that is missed when businesses take a narrow focus on alignment, namely that underperformance is often rooted not just in misalignment but *in the complexity of systems, applications, and other infrastructures.* Together with IBM's *Capitalizing on Complexity* report [2] their primary recommendation to emphasize simplicity is part of a growing realization that to address the dynamism within today's enterprise we have to contain and manage complexity.

Yet simplification is not easily achieved, and the earlier discussion on the enterprise pattern has argued that intrinsic properties of the enterprise include its complexity and system-level behaviors. While all efforts to reduce complexity are laudable, the answers often require the replacement of older legacy solutions with newly architected and more integrated ones which, while no silver bullet, can do a better job of encapsulating the inherent complexity. However, unless new business capabilities are on offer, such IT-centric investment decisions are precisely the type of initiatives that would struggle to make a convincing business case in a business-aligned enterprise. Some timeless, front-end loading suggestions to increase architectural simplicity include the implementation of shared, enterprise-wide standards along with the adoption of enterprise systems with minimal customization. Similarly, newer technologies like virtualization, service-oriented architecture (SOA) and Internet cloud computing all promise to simplify enterprise architectures, yet all require significant levels of investment in time, effort, and management resolve.

Similarly, rightsourcing, another recommended means of escaping the alignment trap, introduces its own management problems. Rightsourcing involves picking the appropriate in-house services to transfer to more competitive and competent external suppliers. Yet outsourcing such nonstrategic business processes and supporting functions poses a management challenge as the enterprise becomes increasingly distributed and virtual. This can serve to exacerbate future change initiatives as they become subject to contractual constraints and renegotiation. Pursuing efficiencies and the commercial benefits offered by a globally competitive marketplace can be attractive but must be traded off against the challenge of managing a virtualized enterprise pattern. Such sophisticated global supply chains require management systems to ensure they are sustainable and do not risk longer-term competitive advantage.

The final recommendation, which the authors' considered a seemingly obvious principle, was to establish end-to-end accountability for initiatives, but this was found to be more the exception than the rule. Their survey found that 74% of IT projects were canceled or failed to deliver the expected results on time and on budget. In an acknowledgment of the gap between strategic planning and execution, true accountability was found to lie in the closer involvement of executives who needed better reporting to be able to measure progress. Those charged with execution likewise need to get the required resources and work closely with the leadership to exercise joint supervision over the projects. While the focus of it also remains on "individual initiatives," it also presents the need for a cultural change towards treating IT as a shared, enterprise resource. These practices reflect the enterprise release planning discussed in Chapter 5, with governance structures bringing together the different businesses that share a common core system.

With a worldwide sample of over 500 business and technology executives, the Bain study provides clear evidence of the disconnect between strategic planning and execution—and of the potential benefit of bridging it. The analysis of their results shows an uncanny similarity to our discussion on change and architectural complexity that highlights many of the common problems to be found in the modern enterprise. The existence of the enterprise pattern is clearly evident in the problems described by survey participants. What is also clear is that the remedies undertaken by leading organizations resonate with key aspects of the release-centric approach. Examples of establishing a multiyear release roadmap for shared systems and the establishment and adherence to a strict policy of four releases a year all point to the central role that release-centric thinking can play in implementing end-to-end accountability, and more generally, in realizing the benefit that arrives from mastering the fourth quadrant.

6.1.2 Downstream Release Governance

The disconnect between strategy and execution is a cause of much frustration in today's enterprises with, mission statements and strategic plans disassociated from the reality on the ground. The alignment trap, and the remedies to avoid it, raise important questions about the style of governance needed to balance alignment and efficacy; agility and continuity. While it is desirable to have seamless and effective, end-to-end governance and accountability, when it is absent or ineffective there is a default tendency to at least establish bottom-up governance mechanisms to fill the void. While not an ultimate solution, these localized efforts attempt to establish protocols from a bottom-up, supply-side perspective that offer local control over key enterprise assets; most obviously IT systems but including areas as diverse as the corporate brand, customer data, and access to facilities. As discussed in Chapter 1, release management has historically been associated with CCM and now subsumes some of those functions in projects. Working with testers and quality control, its current function can be vary in different organizations, but essentially insulating the live, or production environment that supports current business operations from the more volatile development environments. The variability in release management illustrates that while there is an evident need for the discipline, it continues to evolve, slowly, offering the opportunity to bridge the divide between project (and from an enterprise perspective, portfolio) planning with execution and delivery.

6.1.2.1 Guarding Business Operations

Undeniably, governance is most effective early in the life cycle, at the point of project inception and then, as discussed in the previous chapter, at portfolio release planning and design. It is in this early planning phase that the business demand for change can be reconciled with resource capacity and logical interdependencies. Failing that however, supply-side governance emerges to manage the demand for resources and services from a steady train of business initiatives that have the potential to disrupt business operations. In terms of Figure 6.1, these "late and low" efforts reside in the lower right quadrant—later in the life cycle and undertaken at a lower management level.

Change control applied as a defensive measure has a long tradition of pushing back on demand—or accommodating it at a premium. In the halcyon days of the early 1960s when industry and military standards guided, if not directed, early large-scale systems engineering programs, managers would boast that the role of CCM was in fact to "slow down changes" so as not to disrupt agreed development plans. While such a claim would not be acceptable in the modern enterprise, it is nonetheless a legitimate response, at least from a plan-driven, systems-thinking perspective if not a business one—although no one wants to be responsible for "breaking production" and interrupting business operations. Continual changes can interfere with the delivery of a stable and coherent solution, introducing risk which, as a result, increases the cost to the customer. Neither of these perspectives is right or wrong, they are simply twin facets of working with a dynamic system—the yin and yang of solution delivery.

Where strategic initiatives are independently proposed and planned by senior executives without due consideration of enterprise impacts or the organization's execution capabilities, there is an inevitable bottom-up response from those charged with delivery. This can be seen as professionalism, or self-preservation, with some groups in the enterprise establishing protocols to protect themselves from the onslaught of changes. This arises from a need to manage user expectations and avoid disappointing one's customer, particularly in a competitive market where people will do anything to make the sale. Examples of shared functions that apply localized change governance can be found throughout the enterprise. And high-demand functions such as IT, which consistently face a barrage of changes, have made such supplier-side governance an integral part of industry best practices.

The ITIL and COBIT frameworks call for the establishment of change controls and release management (as applied in the latter execution and delivery stages of a project) as core competencies of IT management.The closer one is to the IT production environment the more critical these processes become as they serve to protect the "engine room" of modern enterprise. At this point, protecting core IT services that support the business is a shared priority. Neither business nor IT managers want to have the systems go down and cause business interruption. Increasingly such outages are more than inconveniences to the business—they can be public relations disasters. In the financial services industry for example, IT service outages can directly impact the customer base causing tremendous inconvenience and disruption (as when the ATM network goes down or fund transfers get corrupted) that in turn damages customer confidence and loyalty. So here, downstream in the change life cycle, the deployment of any changes into production have to be rigorously controlled. In this bastion of best practices, the establishment of control processes designed to protect the applications, information, and systems and network infrastructure is not often challenged, even as some of their protocols can at times seem draconian.

It is valuable to contrast these protocols with what takes place in the open market, where, in the face of uncontrolled market forces, suppliers have to establish similar commercial engagement mechanisms. Setting clear and realistic customer expectations in advance of a transaction is fundamental to effective commerce in any marketplace. From simple transactions on eBay to the negotiation of major contracts, success lies in protecting both parties and establishing the terms and conditions under which the goods or services are to be provided. With the increase in outsourcing business support functions, external service providers often have the necessary protocols formalized in contracts, while clients capture their expectations in SLAs that have financial penalties associated with a failure to meet them.

Mature businesses working with multiple suppliers recognize that the logistics associated with procurement and on-time delivery determines the efficiency of their supply chain. Manufacturing companies seek to create close partnerships with their suppliers and prearrange delivery schedules to enable sophisticated just-in-time and lean production strategies. Yet it is within the enterprise boundaries that the greatest inefficiencies can take place. Here, informal processes and apparently cooperating teams can in fact disguise office politics and what, if exposed, would be glaring and embarrassing inefficiencies. At the end of the day however these failings, as shown in the alignment trap, have an impact on the bottom line.

The inefficiencies within many enterprises would not be tolerated if the interactions between internal customers and in-house suppliers were made visible. Where supporting and shared services are within the organization there can be difficulty determining the actual costs, and budgetary manipulation can be made to produce the desired financial results without necessarily delivering any tangible change to productivity or efficiency.

Many enterprises look to solve their problems by outsourcing problematic functions like IT, hoping that it will bring new expertise, reduce costs, and help establish order. The common disappointment with outsourcing after the honeymoon period can be attributed to the fact that changing the players does not change the game.

Introducing greater governance on the supply side without establishing a corresponding linkage with the early planning stages of a change portfolio does little to satisfy the expectations of the business. Indeed, poorly crafted contractual relationships between customer and supplier can increase separateness and damage relationships. As each party looks to defend its own commercial interests, change control mechanisms can be applied as in earlier systems engineering programs to be a contractual tool that quickly becomes the antithesis of agility.

6.1.2.2 Partnering for Agility

The critical role played by IT in many enterprises is indicative of the essential services it offers different parts of the business. As a point of convergence for most change initiatives in the enterprise, managing the demand and supply of its own personnel, as well as those of suppliers and subcontractors is essential while regulate its workload. Therefore, managing IT services is currently one of the more significant challenges in the enterprise. Yet, the well-intentioned IT governance processes can serve to increase costs and create delays and bottlenecks—all in the name of better servicing its customers.

Instead of working as partners with business customers, IT's processes can inadvertently create divisions and be seen as barriers to business growth. Adherence to standard protocols can appear to be unsupportive of strategic initiatives when for example they demand certain preconditions be met. Projects attempting to deliver priority change initiatives can be frustrated by the need to adhere to IT governance processes. Even while they are understood to be essential at some level, they can appear excessively bureaucratic and costly management overheads. Unwelcome governance may be seen to hinder the enterprise's ability to adapt to market conditions. This then is the other side of alignment trap, with attempts by IT to be efficient can appear to make it unresponsive and self-serving.

Essentially defensive, supply-side governance alone can be problematic as each group focuses only on its own unique needs rather than the delivery of the end result, the enterprise release. What is needed is a means of collapsing the services supply chain so it affords greater transparency, collaboration and responsiveness—the very same things that agile methods have attempted to instill within the development life cycle.

This could be considered as process improvement or process reengineering but that misses the underlying cultural change that is required. At a very human level, achieving the connectedness requires not process but psychology, with the alignment and streamlining of the supply chain requiring deep trust between the groups. The different participants in the supply chain have to stop relying just on contracts and governance and instead develop trust and cooperation. The dependencies evident in the enterprise and in its supply chain must create a correspondingly stable ecosystem of customer-supplier relationships as evident in the enterprise pattern.

To deliver strategy calls for the tight coupling of all aspects of the enterprise, the different teams that support it, and the suppliers who participate in the extended supply chain. The alignment of business goals combined with execution competency has to propagate through this complex, dynamic organism requiring it to act swiftly and as one in the face of relentless change.

To achieve this requires more than simply a directive by the board or the CEO. It requires the active and committed participation of all parties—internal or external suppliers contributing to the adjustment of the enterprise. Leadership can drive the direction and be vital to inspire others, but to actually execute strategy requires coordinated action.

In seeking a more balanced governance model, one misconception needs to be put to rest. Top-down governance was never intended to be interpreted as relating to the organizational hierarchy. It was always intended, from its early uses in systems design, to be related to a big-picture, architectural perspective—in the context of this book, what we have termed an enterprise perspective.

Successful businesses are conducted by a multidisciplinary team working closely together for the benefit of shareholders. The essence of innovation is that strategic decisions should not emanate from the CEO or executive leadership alone. When crucial trade-off decisions are made about strategic direction, individuals with specialist knowledge should be actively involved—not just represented by management. The combination of management responsibility with specialist knowledge is a key motivation for agile methods which, as discussed in Chapter 2, attempts to minimize management overheads and enable experts to have a dialog directly with the customer.

Being a specialist or technologist does not disqualify one from being a strategic thinker and it is not only executive management who necessarily have the best interests of the business at heart. Indeed, with aggressive performance bonuses that focus on short-term business goals, it may be that the more reliable and sustainable strategies would emerge from staff with a longer tenure in the organization. While such directives and associated reward structures must come from the CEO or even board-level, clearly governance processes can help ensure that the longer-term interests of the shareholders are protected from potentially disastrous strategies.

Business needs to return to being a collaborative endeavor where considered and rational debate can be had by different and sometimes opposing parties. Partnership is paramount to establishing such an egalitarian governance model and needs visibility of the strategic change portfolio and participation of key suppliers in planning at the enterprise level. Agility is required not at the project level but at the portfolio and enterprise release level.

In an environment where projects have become business as usual and embedded in the organizational hierarchy, a higher-level independent governance body becomes necessary. Avoiding the alignment trap calls for the establishment of an enterprise-level multidisciplinary governance framework that encompasses architecture and strategic sourcing as well as business prioritization. While such a framework may support the traditional project paradigm, applying the proposed remedies effectively requires the adoption of an enterprise-level view of delivery that results in a profound shift in the power and influence of a project.

Emphasizing simplicity at the enterprise level constrains a project's solution design to work within the supported features of the chosen enterprise system or comply with strict architectural standards. Rightsourcing can restrict the range of suppliers with whom a project can negotiate the implementation or support of the solution. Finally, end-to-end governance will necessarily restrict any flexibility the project has to redesign or replan after it receives funding resulting in the need to undertake more comprehensive design and estimation up front.

6.2 Assessing Delivery Patterns

Words like collaboration and partnering are somewhat meaningless and little more than platitudes unless we know the purpose to which they apply. To attain a more integrated governance with the end-to-end accountabilities the Bain Study recommends, we need to understand the context and be more specific about the roles that different parties play in delivering change in the enterprise.

As previously discussed, the very definition of what constitutes a project has been subtly altered over time in response to continuous change. And with the enterprise pattern, it becomes clearer why the very focus offered by projects can limit its ability to address enterprise complexity—indeed, a project's narrow viewpoint is more likely to exacerbate complexity.

The assertion this book makes, and herein sets out to establish, is that that there are fundamental forces at work that require us to instead adopt a release-centric perspective to delivery. The packaging of what is being delivered needs to be reassessed and move beyond being a random collection of changes to become a planned and designed delivery of the next version of the enterprise. Increasingly, the appropriate paradigm around which to build such an execution competency is in fact a release, not the project.

To support this contention, this section considers the customer–supplier relationship that is fundamental to addressing the demand and supply equation that has the potential to derail the best intentioned change programs. The previously introduced enterprise pattern of Chapter 3 will be shown to be the most complex instance of such a relationship which has emerged. Considering the different relationship types possible, irrespective of whether these are external contractual agreements or in-house commitments, provides a convenient heuristic for determining the appropriate management response. As these delivery patterns, as we have termed them, emerged, they required an equivalent adjustment to our management patterns. And as the simple analysis highlights, there are clear limits to the effectiveness of the project paradigm that can struggle when the assumption of control breaks down.

6.2.1 The Customer–Supplier Relationship

By considering the customer-supplier relationship we attempt to separate the form of the execution and delivery problem from its prospective solutions. Under the prevailing project paradigm, separating the "what" from the "how" is no easy task when we have become accustomed to discussing the two interchangeably: projects are the accepted delivery mechanism for satisfying any reasonably complex customer-supplier agreement that is not simply purchasing commodity items.

The customer–supplier relationship is fundamental to all commerce, and therefore is at the heart of execution and delivery management practices and methods. The commercial transaction between two parties is the basis of any contract, formal or implicit, and has been the guiding principle at the heart of what determines successful delivery. Suppliers must "delight their customers" and if they cannot do that they must at least meet the customer's requirements, if not always their expectations.

In analyzing customer–supplier patterns we attempt to avoid, as much as possible, solutioning language. While acknowledging that describing a problem in familiar terms can help understanding and be useful communication shorthand, this convenience needs to be used sparingly when we are attempting to study patterns and differentiate alternative approaches. So, in what is to follow, we attempt to keep our language precise to avoid potential confusion, only applying loaded terms such as project and release when interpreting the patterns that emerge.

6.2.1.1 Evolving Delivery Patterns

The relationship between customer and supplier is pervasive and is relevant across different business and technology domains. It is central to quality management as defined by international standards including ISO-9000, and best practice frameworks such as the CMMI and PM-BOK. These aim to ensure that key aspects of the customer–supplier relationship and interactions are defined to reduce the risk of misunderstandings and nondelivery.

The formality of the relationship can vary based upon the manner in which the customer's requirements are relayed and its delivery tracked, as well as, of course, the consequences of nondelivery. Such relationships can include internal agreements between teams and individuals who perform key activities in the delivery life cycle. With this interpretation, we can consider customers and/or suppliers to be internal or external entities, either individuals or external agencies that are engaged in the delivery process. Delivery itself can be at any point in an extended supply chain where there may be numerous handoffs between parties until the end customer receives the required solution.

The Agile Manifesto can be seen to take a demand-side, customer perspective with one of its four central tenets calling for "customer collaboration over contract-negotiations." While traditional, plan-driven approaches apply a supplier perspective which, as previously discussed, takes a more contractual view of what constitutes successful delivery. The informality of the former constitutes an acceptance of risk by both parties, while the latter attempts to reduce that risk by providing detailed requirements and a more exacting specification of the customer's expectations. In either case an agreement must be struck between both parties that establishes the scope of what needs to be delivered.

At the very heart of all mainstream delivery, best practices is the representation of this two-party interaction between customer and supplier. Yet the implicit assumption has been made that the nature of the customer–supplier relationship has been static over recent times. The enterprise pattern showed unique properties because it broke that fundamental assumption, and in the process showed a closer correspondence to the reality experienced in the enterprise. But there are intermediate stages in the customer–supplier relationship that can offer a historical perspective and context for the different management patterns and approaches that are evident in organizations today. By figuratively dialing up change and complexity incrementally, we get a better appreciation of evolving delivery patterns and the management challenge they represent, culminating in the enterprise pattern previously discussed.

The generic and fundamental relationship between customers and suppliers provides a means of investigating the delivery challenge as the enterprise grows increasingly integrated and interdependent. The simple analysis undertaken below illustrates the evolving pattern that acts as a a visualization of the increased change and complexity managers have to contend with. It further serves to highlight that the assumption of a two-party customer–supplier relationship can seem rather quaint in the globalized economy and the complex supply chains it offers—including but not limited to the now common outsourcing or offshoring options available to businesses.

Figure 6.3 shows each of the four basic relationships possible between customers and suppliers, using simple data modelling notation. Essentially, a single customer may engage a supplier to provide the needed solution, or where the requirements are complex in nature, engage several teams or subcontractors who provide separate solution components that need to be integrated. Alternatively, the customer may be one of many, in either the market or the enterprise, who either works directly with one supplier or multiple component suppliers.

In the discussion that follows, and as a concession to the prevailing project paradigm, a program or project is considered as the default delivery vehicle for each customer's requirements. A supplier can be considered to be responsible for the delivery of an entire solution, or a solution component, that addresses the needs of the customer. While this language will be adopted as it helps in the discussion of the different delivery patterns, it is only one interpretation. Depending on the nature of the delivery, whether it is a business change, the

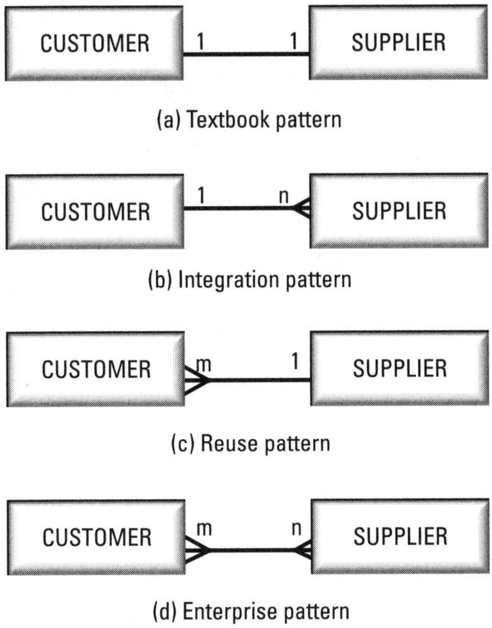

Figure 6.3 (a–d) Alternative delivery patterns.

development of an IT solution, a consumer product or service, the language can be adjusted. A more visual matrix representation of these patterns will be described in Chapter 7 and supports a generalized release-centric paradigm that offers a new perspective for strategic execution in the enterprise.

Each of the delivery patterns is discussed below to help clarify the underlying assumptions they make about managing execution and delivery. These scenarios make us appreciate the circumstances under which corresponding management approaches are appropriate. Each delivery pattern offers an insight into the effects of increasingly complex enterprise architectures and the limitations of current project-centric thinking. It also shows that a release-centric approach is not always necessary and managers therefore have the option of applying the management approach best suited to their environment.

6.2.1.2 One-to-One: Textbook Pattern

The simplest, idealized relationship between customer and supplier is the unique, one-to-one relationship. It can be considered the textbook pattern because a single project is established to provide a solution delivered by a sole supplier. It is the original two-party scenario that is most prevalent in management literature and being the easiest to describe, forms the basis for most of the standard project management terminology.

In the textbook delivery pattern shown in Figure 6.3(a), the customer can exert full control over the single supplier responsible for the solution whether that supplier is external or in-house. Usually this control is exercised financially for external suppliers through payment milestones linked to measures of progress or delivery of key functionality. Suppliers internal to the enterprise are "motivated" through group and individual objectives and performance measures. These ultimately also have a financial basis, if not through bonuses and pay raises then through the ultimate incentive of keeping one's job.

The one-to-one relationship is historically the earliest type of interaction between customer and supplier. It can be seen to be representative of greenfields or bespoke projects, where both parties work together on the development of a solution. Such projects can sometimes be more about understanding new capabilities and exploring their possibilities and so are amenable to agile methods and their adaptive, iterative release cycles. The relationship may of course persist as the solution enters the maintenance phase or be terminated as a new supplier is engaged.

The textbook pattern gives rise to familiar project management terminology and its delivery life cycle, undertaken within the customer's constraints of cost, schedule, and quality. All the usual activities that need to be undertaken in the solution delivery life cycle have to be conducted and do not warrant repeating here. The methods and tools adopted as a part of the solution life cycle are immaterial to the broader relationship between the two parties—the project may be either agile or plan-driven. But what is important is that there is clearly control exerted by the customer, albeit via the project manager or other proxies, which in its most basic form consists of some manner of financial arrangement between the two parties.

While managing projects satisfying the single customer-supplier pattern is well understood and documented, the delivery of such projects is not trivial. Even as the problems encountered in the textbook pattern have preoccupied the management community for decades, the delivery challenge remains relevant to many managers today. However, as we proceed with our analysis, we will see that this simple pattern, which allows the greatest management control to be exerted, is also the pattern for which most management tools have been developed.

6.2.1.3 One-to-Many: Integration Pattern

One-to-many customer-supplier relationships [Figure 6.3(b)] give rise to what we term the integration pattern and describes a more complicated delivery pattern. Here a customer has to rely on multiple suppliers, each providing a component of the solution architecture. While there are several scenarios that could give rise to this pattern, its defining feature is that the different solution components eventually have to be integrated to deliver the complete solution.

From a historical perspective, this is a familiar pattern that would be traditionally managed as a program. The pattern is often evident at a fine-grained level whenever a solution requires the contribution of several specializations. The integration pattern is not confined simply to technology architectures and can be seen as a consequence of most architectural techniques, from component-based design and the more recent SOA, to organizational design and business process architectures. The pattern emerges naturally, and as a solution architecture scales, specialized solution components emerge that may be supplied by in-house departments or procured externally from vendors.

The emergence of solution components can be an indication of a maturing market, as increasing specialization gives rise to suppliers developing specific point-solutions with advanced capabilities. Aside from the increased quality, functionality, and reliability these solution components might offer, specialist suppliers can focus on increasing the efficiency that allows competitive pricing.

Customers can benefit from utilizing these specialist suppliers so long as they are prepared to adopt the integration pattern. Increased management overheads and integration risk has to be traded off against the advantages of higher quality components with better price-performance ratio. The adoption of such a "best of breed" strategy requires the customer to integrate solution components from different suppliers together to meet their specific needs, rather than source the complete solution from a single supplier.

Mergers and acquisitions can also give rise to the integration pattern when multiple, disparate, and sometime duplicated solution components have to be integrated together to provide a uniform customer experience. While the degree of the challenge in reestablishing a new combined enterprise architecture is not always understood at the outset of any merger or acquisition, it is a direct consequence of a delivering more sophisticated, competitive, and integrated products and services to the market.

Irrespective of the manner in which an enterprise arises at the integration pattern, they share similar challenges in the delivery, support, and evolution of their particular configuration of solution components. The integration of these components into the desired enterprise solution can be a significant challenge in itself, one that has to be addressed not simply at the initial delivery but maintained continually as new releases are offered by the suppliers. It implicitly assumes a design authority because rarely is a solution simply a collection of components. More likely there is the need to define compatible interfaces, common design principles, and for the components to adhere to agreed standards. The creation of an integrated whole is never easy and disciplines such as systems engineering exist to establish and maintain a solution's design and architectural integrity through its operational life.

Instead of managing the complex program that would be needed to manage the multiple suppliers and integrate the solution components, the customer has the option of avoiding these challenges by assigning a system integrator or prime contractor. These outsourced management intermediaries offer the integration expertise to coordinate multiple subcontracted and in-house suppliers and give the customer a single point of contact for the solution delivery. In these instances, while the customer may have simplified the management interface, there is no net reduction in the delivery challenge; it has simply been shifted to become someone else's responsibility. As in the second law of thermodynamics, which states that entropy can only increase or remain constant, there is no avoiding complexity. The risk may have reduced and the likelihood of delivery success improved due to the system integrator's greater management experience, but it could also be likely that new risks are introduced by the introduction of more personnel and yet another contractual boundary.

6.2.1.4 Many-to-One: Reuse Pattern

The many-to-one customer-supplier relationship is termed the reuse pattern [Figure 6.4(c)] and can be seen as a result of specialization and maturation of a market that sees the emergence of internally reused components or commercially packaged solutions to common problems. While internal reuse of software and business components remains an elusive challenge, the growing popularity of commercial software packages provides evidence of the strength of this pattern. Wherever there is sufficient demand for a particular solution (or more generally, a solution component) the efficiencies resulting from the equivalent of mass production can establish a viable market and give rise to this pattern.

From the supplier's perspective, achieving the reuse pattern represents the pinnacle of success as what might have started as a project matures into a product or component. Where internal to the enterprise, the reuse of common or shared components presents the opportunity to eliminate duplication of effort and consolidate systems and processes. For vendors it represents a successfully marketed and productized solution component that has met the needs of its customer base. The supplier's challenge lies in responding to the different requirements customers, and prospective customers, have for the solution, and pacifying those that do not get the features they want. As in managing any customer base, the art of compromise is needed to operate within the reuse pattern. While there are a multitude of vendors in the market, the lack of control that in-house suppliers have over their customers means there are far fewer examples of systematic component reuse within organizations.

The customer's decision to purchase commercially available solutions offered by specialist vendors is often a low-risk option for an enterprise when compared to the uncertain costs of developing and maintaining equivalent in-house solutions. They benefit from the amortization of overheads across multiple customers having the potential to reduce costs while improving quality and offering the prospect of more frequent upgrades with new capabilities. The success of commercial enterprise systems packages offered by suppliers like SAP and Oracle attests to the fact that many enterprises have similar requirements that can be met by a common solution, their uniqueness restricted to configuration settings and business rules.

There are also drawbacks associated with the reuse pattern, with customers now receiving a generic solution which may not perfectly match their needs and cannot provide a point of differentiation against its rivals that could also have access to the solution. Further, specific enhancements and upgrade requests are no longer within the customer's control and arrive, if at all, at a point determined by the supplier. Finally, the frequency and timing of enhancements may not suit the enterprise, and yet they may be forced into an upgrade cycle by the supplier.

Certainly the management implications of the reuse pattern are quite profound, as it represents switching control from the customer to the supplier. Customers now form a collective and no single customer can directly influence the direction of the solution or demand features that diverge from the interests of the collective. Control over the solution's evolution has been effectively traded for anticipated improvements in quality, expanded capability, and reduced support costs. As such, the management challenges evident in the reuse pattern provide the first signs of the erosion of the traditional command and control management approach. In reality, customers have some negotiating power before the solution is purchased, but unless they represent a major proportion of the supplier's revenues, their influence on the solution's future direction is often short-lived.

6.2.1.5 Many-to-Many: Enterprise Pattern

The many-to-many customer-supplier relationship is the enterprise pattern [Figure 6.3(d)], a generalization that encompasses all the other patterns, and was introduced in Chapter 3. The key feature of the pattern is the manifold relationships between customers and suppliers that in the enterprise can be attributed to the sponsors who place demands on functional groups like IT. Each customer may rely on several internal or external suppliers, while these suppliers provide components that are utilized by different customers.

The pattern can be seen to represent a multiplicity of customers sponsoring concurrent projects or change initiatives that impact an integrated enterprise architecture. The suppliers responsible for the shared solution components may be in-house or part of a globalized marketplace as a result of outsourcing, offshoring, or old-fashioned subcontracting. The enterprise pattern therefore represents the modern, virtual enterprise where customers (or more accurately their in-house proxies) rely on the offerings of multiple supplier teams, each responsible for specific solution components that have to be integrated together with other components to provide the desired solution or capability.

The enterprise pattern captures the behavior of a system of systems or a federated structure where the participants behave autonomously and there is no single point of control or coordination. Where this occurs with commercial solutions and external suppliers, there is little distinction between this pattern and the reuse pattern; it simply represents a wider marketplace where customers can purchase different solution components from several vendors. In both patterns, the supplier has the responsibility to keep their customers happy and that means their focus needs to be on robust quality products, frequent upgrades, and good user support. However, where the pattern is evident within the bounds of a single enterprise, the existence of multiple customers, each with their separate requirements needing to be satisfied concurrently, makes for a new and unique management scenario. In such a scenario, the supplier of a shared solution component cannot find an authority that can arbitrate the priority of conflicting or contentious requirements placed by different customers.

The critical point argued in Chapter 3 is that the enterprise pattern describes the situation found in many large enterprises and that can cause traditional project management approaches to fail. Management control is the underlying assumption behind the prevailing project paradigm, and it breaks down when the influence a customer has over suppliers is diluted. A supplier with the ability to provide their services to other customers essentially has greater autonomy and negotiating power, undermining a customer's ability to dictate terms. Further, in all preceding patterns one party has always had control over the customer–supplier interaction, even though that was the supplier in the reuse pattern. The peer-to-peer nature of the enterprise pattern therefore presents a unique management challenge. Not only is the customer without full control, there is no one party that can claim to have any control. Power is fragmented in this egalitarian pattern; all customers can be seen to be peers and the only means of overseeing the entire pattern is to establish overarching governance processes across the individual transactions.

The lack of a central controlling agency differentiates the enterprise pattern from all previous delivery patterns. The combined effect of the mutual benefits and interdependence of the customer–supplier relationship and lack of direct control necessitates all parties work cooperatively and collaborate to achieve the desired outcomes. Unlike the textbook pattern where the project paradigm has reached near universal acceptance, the appropriate terminology to describe the enterprise pattern remains to be formalized.

What is clear, however, is that the pattern corresponds to that of a portfolio of interdependent projects making demands on shared component suppliers. In that scenario the pattern illustrates the need for release-centric governance to coordinate the separate activities and reduce, as far as possible, interference and execution risk. The separate projects must participate in some level of shared planning and execution to preempt integration problems at delivery milestones. How this is achieved can vary. A high-governance culture can demand that approved projects undergo a shared release planning and design process that arrives at a release architecture and roadmap, as discussed in the previous chapter. Alternatively, a less formal and more collaborative approach can see the codesign of shared components without the creation of a new layer of management—arguably what happens today in the absence of a release framework. In either instance, project dependencies see the need for greater sharing of project plans and architectures across the life cycle. And in order for the necessary crosstalk between projects not to become a burden, new release-centric processes with succinct representations will be valuable for improving collaboration *between* projects—something that has not been as critical in the past.

6.2.1.6 Management Complexity in the Enterprise

The simple analysis described in this chapter identifies four stages of increasing management complexity with each stage representing a particular business environment and culture. Together these stages provide a framework for discussing the appropriateness of different management techniques.

Figure 6.4 presents the different management patterns overlaid on the four change-complexity quadrants. The correspondence between change and customers justifies the vertical change axis in the figure. Where there are more customers, there can be expected multiple streams of change (or projects) that sees the reuse and enterprise patterns, with their multiple customers being associated with greater change. Similarly, the horizontal complexity axis relates to the number of suppliers (and so components) that have to be coordinated, integrated, tested, and released. Hence the integration and enterprise pattern can be said to have greater complexity as they must grapple with the integration of multiple suppliers' components.

Both grids can be seen to be a result of change and complexity taking different forms. In relating Figure 6.4 to the previous Figure 2.3 we can see which management approaches are most suited to the delivery patterns. Further, synthesizing the finding of the Bain study summarized in Figure 6.2, we can add the dimension of alignment versus efficacy. The correspondence suggests that alignment and agility can be expected to operate well under both the textbook and reuse patterns, so that simple developments or procured products will not constrain the business as it attempts to quickly adapt solutions to its needs. Standardized practices and planning work well to address the textbook and the integration patterns. Again, the textbook pattern is simple enough that any approach can be made to work equally well. But it is the fourth quadrant and the enterprise pattern that is clearly the most complex and poorly addressed—a gap in our management arsenal. While this is not to suggest that the enterprise

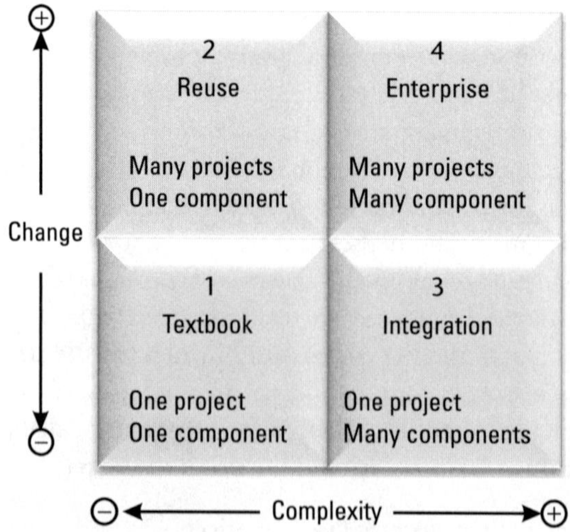

Figure 6.4 Alignment versus efficacy .

pattern is the sole cause of the complexity found in the enterprise, it does capture an important facet of the challenges relevant to the strategic delivery of change.

This fourth quadrant has also been shown to be problematic in evolving product line architectures [3] that also satisfy the enterprise pattern. Increasingly common in software and consumer electronics, where multiple products share a core architecture, or platform, they present significant management challenges that have been described as "stretching" traditional management practices [4]. Similarly, a radical reassessment of management approaches in the enterprise is needed. And even as the enterprise pattern represents the most complex management scenario, recognizing it allows new models and representations better suited to address the enterprise's complex reality.

References

[1] Shpilberg, D., et al., "Avoiding the Alignment Trap in IT," *MIT Sloan Management Review,* Vol. 49, Fall 2007.

[2] Institute of Business Value, *Capitalizing on Complexity,* IBM Corporation, 2010.

[3] Taborda, L. J., "Managing Change and Architectural Complexity: An Integrative Approch," Macquarie University, Ph.D. Thesis, 2005.

[4] Clements, P., and L. Northrop, *Software Product Lines: Practices and Patterns.* Reading, MA: Addison Wesley, 2002.

7

The Enterprise Release Paradigm

... by relieving the brain of all unnecessary work, a good notation sets it free to concentrate on more advanced problems and, in effect, increases the mental powers of the race.
—Alfred North Whitehead, mathematician and philosopher

The previous analysis examined how management challenges evolve as complexity grows and organizations became more segmented and specialized. The different delivery patterns that have arisen as a result culminate in the enterprise pattern, a generalization that encompasses the other, simpler patterns and embodies the management challenges foreshadowed in the fourth quadrant of Figure 2.3. The project paradigm cannot adequately control and coordinate changes emanating from multiple independent sources in the enterprise. And the trend towards consolidating and sharing business and technology components only exacerbates the communication and negotiation necessary to achieve a strategic result.

This chapter attempts to tackle the root cause of the problems evident in the enterprise pattern without preconceptions and avoiding the entrenched project-mindset. The groundwork laid in preceding chapters now comes to a resolution with the generalization of the traditional release construct that directly addresses the multiplicity of dependencies encountered in the enterprise. Developing the argument from first principles, we derive a simple representation of an enterprise release that we term the release matrix.

We show that the matrix notation arises naturally from the multiplicity and provides a concise means of capturing the interrelated customer–supplier relationships that culminate in an enterprise release. As a result, this representation provides a unifying perspective of the changes occurring in an enterprise at the abstraction level needed to effectively communicate with and coordinate the different parties. The underlying PM fundamentals remain relevant, albeit at a more detailed level—now with a better appreciation of the dependencies between the apparently separate initiatives.

We will see that a release matrix acts as an aggregator of the individual project information, providing executive and senior management with a more holistic enterprise view. It provides the basis of simple tools that apply equally to the planning and execution of a portfolio, enabling an end-to-end enterprise release life cycle to be defined. With the matrix notation a representation of the change portfolio becomes possible, one that introduces new dimensions into the discourse and makes tangible the enterprise release paradigm.

7.1 Generalizing the Release

The ascendency of the release construct has resulted in a growing recognition that it offers a shared, end-to-end delivery perspective that can be understood by customer and supplier alike. The release has relevance at the execution and delivery end of the life cycle, but is equally important for setting expectations in the earlier stages of release planning and design.

The release is a neutral concept, a touchstone and common goal, for both the business as well as enabling services like IT. It is central to agile methods with regular, short sprints and continuous integration bringing the realization that managing the release from the outset is critical to both speed and quality. Prioritization and release planning of new features with the customer is an important first step that can smooth the development path. Similarly, the wider IT services management community has embraced the release in their ITIL and COBIT management frameworks. And we should not forget that discussions about releases and roadmaps are routine in product management, irrespective of the nature of the product itself.

However, with the increasingly wide usage of the term, there arises a need for greater genericity in the release management discipline. Beyond simply recognizing the release as the ultimate vehicle for delivering incremental change, there is a need to extend our language to address the concurrency of multiple projects we find in the enterprise. When requirements scale to entire projects and even programs, the release construct must be generalized to accommodate the selection, planning, and execution of an enterprise change portfolio. At that point it becomes clear that the release is not simply a synonym for project—it becomes a collective noun for the delivery of an entire portfolio.

As evident in the enterprise pattern, the number of sponsors, stakeholders, functional teams, and suppliers participating in a release can give rise to new coordination challenges. Each sponsor (customer) and functional team (supplier) can view the release through their own unique lens. The multidisciplinary nature of enterprise releases, with their supply chains extending across both internal departments and external business partners, results in the overloaded and ambiguous release management terminology. There is a risk that the very forces that give rise to the increased relevance and acceptance of the release also conspire to confuse terminology and concepts.

There has been a tendency for management best practices to focus on process—the planning and coordination of development activities. This was perfectly reasonable given that most development was greenfield and the resulting application, system, or solution was not yet due to come into existence. It was left to those responsible for maintenance and support to address incremental changes and the need to retain the products or systems architectural coherence. But in continually evolving architectures, be they product, solution, or enterprise, these two perspectives—process and product—must come together. The shared recognition of the release and its ability to be the nexus between process and product makes it a unifying construct.

The generalization of an enterprise release introduced below captures the multiple customer and supplier perspectives of an enterprise release in a release matrix. By aggregating the changes sought by customers, it provides a mechanism for understanding and guiding the evolution of components of the enterprise architecture. Further, the release matrix representation is a true generalization in that it reduces to the specific and familiar definitions of a release the result of a generalized customer–supplier relationship that is the enterprise pattern.

7.1.1 Introducing the Release Matrix

The increasingly complex delivery patterns discussed in Chapter 6 give rise to the generalization of an enterprise, or portfolio release: an aggregated and integrated multiproject delivery vehicle. If such consolidated releases have not been previously considered it was simply because they had not been necessary while the textbook and integration patterns were predominant.

The emergence of the complex, multiple customer–supplier relationships described by the enterprise pattern needs to be recognized as earlier management approaches become less effective in dealing with enterprise complexity. These fundamental interactions, with parties both internal and/or external to the organization, become critical when attempting to manage change and transformation. The pattern is evident across a range of industries and independent of the nature of the capabilities being delivered. When the control a customer exerts over its suppliers breaks down, collaboration and negotiation become vital.

7.1.1.1 The Multidimensional Release

The release matrix provides a means of defining and tracking the various agreements and commitments made between the different parties sponsoring changes or responsible for delivering the solution components. This collection of interrelated customer–supplier agreements must be coordinated to satisfy time-to-market demands and maximize the value to the enterprise; that is, deliver on the enterprise strategy. Ensuring that the separate plans and commitments of the suppliers align with that end goal requires a new level of coordination and synchronization which requires we have a consolidated view of the change portfolio—that can itself adapt to changes in business priorities.

It is the lack of a centralized power in the enterprise pattern that results in a devolved, federated management structure where collaboration and negotiation necessarily replace control. These are the conditions that arise in most large organizations where power is fragmented and achieving any strategic outcome at the enterprise level requires gaining the support of numerous sponsors. This is particularly challenging when the business demand exceeds supply-side capacity—a consequence of unrelenting changes placing pressure on core support functions like IT. Getting the desired results in these circumstances requires intense negotiations between the parties involved that aim to align plans for the execution of an agreed change portfolio. In such a portfolio, there is not a single release that needs to be planned and executed; there are as many releases as there are participants. Where there is no overarching enterprise roadmap, planning and negotiating with suppliers independently results in a frustrating free-for-all that consumes precious management-time, corrupts core components of the enterprise architecture, and is rarely aligned with strategy.

The many-to-many relationship that characterizes the enterprise pattern denotes a portfolio comprising multiple projects that impact components of the enterprise architecture. This interpretation of the pattern shown in Figure 6.3 (d) together with simple data analysis techniques for resolving such relationships [1] lead to the introduction of a new associative entity identified in Figure 7.1 as the release matrix. What this denotes is that there are in fact multiple releases taking place in the enterprise pattern. As discussed in Chapter 5, projects may have delivery milestones independent of each other. This places stress

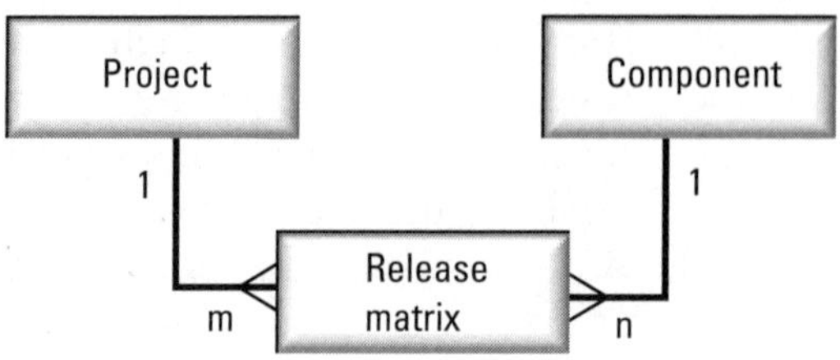

Figure 7.1 Deriving the release matrix.

on those core components which, being critical to multiple projects, are forced into a cycle of continual releases. The resulting overheads such on-demand releases entail can drive up costs as well as introduce quality problems. These factors drive organizations towards more regular, enterprise releases.

What the simple data model serves to highlight is that such a generalized, multi-project release is a unique entity in its own right that arises naturally from the enterprise pattern. It can be seen to play an important role in bridging the dynamic, activity-based project dimension with the more stable and static architectural one. This will be shown to be a crucial feature of the notation: that it makes a clear distinction between the management of the dynamics, the temporal changes impacting the enterprise, and the architectural considerations that make for stability.

On the left of Figure 7.1, we have the transience of the project world, while on the right we have the continuity provided by components of the enterprise architecture—with the release matrix promising to be the bridge between the two perspectives. Both projects and components can be seen to have "entries" in the release matrix that allow their different needs and constraints to be clearly represented. These entries are the individual release schedule a project establishes and which its component teams must match to enable the solution. In the case of an enterprise release, multiple projects and the impacted component teams must agree and synchronize deliveries to meet agreed release milestones. By making visible both sides of the demand–supply equation, the release matrix provides a collaborative tool for negotiating the release scope and cost of each release with impacted component teams. A separate release matrix must be negotiated for each release, as newly initiated projects can arise and must compete for priority with earlier projects for limited resources. This can be an orderly process where release-centric governance and planning framework discussed in Chapter 5 is implemented.

7.1.1.2 Explicit Dependency Management

Current delivery management practices, centered as they are around the project paradigm, are somewhat simplistic and linear, more suited to the singular "text- book" project. They are literally not designed to cope with the cross-project dependencies inherent in the increasingly volatile and complex modern enterprise. Addressing these dependencies and the impact of multiple projects on the enterprise architecture can reduce the confusion and inefficiency often found in the enterprise. Visibility of separately managed projects is essential to make strategic decisions—even more important, to do so in a timely manner and with confidence knowing that key priorities will be met.

Understanding the impacts of multiple changes on the enterprise architecture can go a long way in addressing some of the confusion and inefficiency to be found in the enterprise. The lack of visibility of separately managed changes makes it impossible to make strategic enterprise-level decisions and unclear dependencies can mean that unintended consequences arise.

When presented in its tabular or matrix notation shown in Figure 7.2, the release matrix provides an alternative, compact representation for the multiple, concurrent initiatives captured in the enterprise pattern and discussed back in Section 3.1. As the tabular format makes the naming of the release matrix clearer, it can be seen to offer multiple projects and their component suppliers the ability to share their dependencies and mutual commitments with each other.

In identifying the different projects' impact on component teams, the notation helps clarify the different viewpoints that have to be balanced in planning an enterprise release, the separate demands of the projects versus the capacity of the component suppliers to deliver the required capabilities. The rows of the matrix denote the projects (Pi) that are being executed concurrently while the solution components (Cj) form the columns. When a relationship exists between the project (Pi) and the component (Cj), that cell of the matrix contains a 1, with a 0 denoting that no relationship exists. Therefore, the impact of a project on a component is represented by a line in the illustration that translates to a 1 in the corresponding cell of the matrix. But this concise notation can easily extend to capture more useful information about the details of the relationship between projects and component teams.

7.1.1.3 Multiproject Release Scope

A release matrix provides a consolidated view of an enterprise release, and indeed can serve to define the construct. The matrix is spanned by the project portfolio and architectural component inventory (some might prefer system inventory or even application portfolio). These two vectors establish the scope of a specific release, capturing all participants involved, along with their expectations and responsibilities of each party involved. Thus, the notation represents a snapshot of the decisions made in determining the contents of the release, based on the planning and design described in Chapter 5.

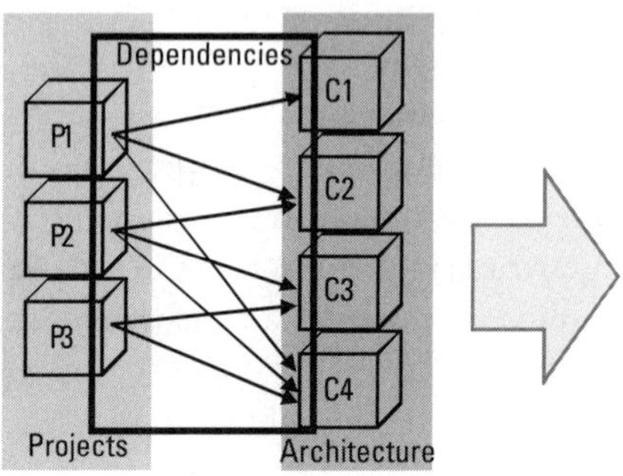

Projects	Architecture			
	C1	C2	C3	C4
P1	1	1	0	1
P2	0	1	1	1
P3	0	0	1	1

Dependencies

Figure 7.2 Enterprise pattern in matrix notation.

By explicitly capturing the project impacts on components of the enterprise architecture, business requirements are separated from the enabling capabilities needed to implement them. In this manner, the matrix notation reinforces the orthogonality of the customer and supplier dimensions and provides a means of exploring the different relationships inherent in the pattern. If the axes of the matrix capture participants in the release, then the tabular body of the matrix captures the detailed relationships between these parties. The intersection between a specific row and column of the release matrix captures the individual agreement between the customer and the supplier.

Figure 7.3 illustrates the use to which the notation can be applied, expanding on the project-component supplier relationship to show that the body of the matrix provides a convenient representation for the requirements that form a release's scope. In the example illustrated, the entire matrix represents a specific release comprising three projects that have committed to deliver capabilities within its timeboxed schedule. The solution to be implemented by project P2 is shown to place requirements on the C2, C3, and C4 component teams. In particular, the highlighted row and column show that the project has to negotiate with the C2 team for delivery of the requirements denoted as r_{22}, the subscript indicating its position in the matrix. This is effectively the contract between the P2 project and the C2 component team, implicit when the parties are within an organization but formal and explicit where C2 is an external supplier.

The matrix representation of a multiproject, multiteam release helps to crystallize earlier discussion of a holistic, bird's-eye perspective of enterprise evolution. It provides a complete specification of an aggregated release from the active participants down to the specifics of the requirements that define the relationship between projects and component teams. The mere creation or existence of a release matrix represents a planned or current release that will have as a minimum a name, a set of integration and test milestones, and a delivery date. At its perimeter, a release matrix defines its broad scope in terms of the projects and component teams participating in the release. This is the release

	C1	C2	C3	C4
P1	r_{11}	r_{12}	0	r_{14}
P2	0	r_{22}	r_{23}	r_{24}
P3	0	0	r_{33}	r_{34}

Figure 7.3 Release matrix for requirements.

team, or more appropriately, the community—those sponsors, stakeholders, departments, or teams (some external agencies) who will be involved in the release. Finally, the body of the release captures the requirements that have to be delivered as one in the defined release window. These represent the business solutions that have to be delivered as an integrated whole, as defined by the release architecture, and with the components providing the enabling capabilities.

7.1.2 Resolving Enterprise Multiplicity

The release matrix can be viewed as a generalization of the familiar project release. As such, it offers a notation suited to multiproject portfolio or enterprise releases, but can be reduced to more familiar project management concepts. Mapping the notation to current management approaches helps both in understanding the implications of the matrix and how it can be usefully applied as a tool for managing enterprise releases.

When considering the rows of the matrix, we take a project perspective that shows the release matrix can be an extension of traditional requirements management practices. Similarly, focusing on each of the rows in the matrix takes the perspective of the different component suppliers who must scope requirements and estimate costs before committing to a delivery schedule. The key to the matrix's usefulness is that both these perspectives have to be balanced as part of planning, with the final agreements locked down in a baselined enterprise release plan.

The participants of each release have their specific goals that may not all be met at once, requiring negotiation and compromise. Returning to Figure 7.3 we can see that the P2 project needs the C2, C3, and C4 component teams to meet its requirements before it can deliver its solution. However, each of the separate component teams has to design for the collective requirements of multiple projects. C2 is a case in point, needing to address requirements from P1 and P2. While they may be quite independent requirements, if the two projects emanated from different parts of the organization, it is possible their requirements are conflicting or at least have design implications that need to be addressed.

7.1.2.1 Requirements Allocation and Traceability

With decomposition and specialization a common response to growing complexity, even relatively simple changes in the enterprise can impact several parts of the architecture. As a result, there is a need to plan, design, and coordinate execution and delivery across several suppliers irrespective of the scale of the change, whether it is a program, project, or a seemingly trivial change.

While less of a concern in our historically siloed enterprises, the distributed impact of change is a recognized and familiar pattern in software and systems development. It is evident at different levels of granularity, from the need to modify several software files to affect a simple change, to the need to coordinate activities across multiple teams, departments, or contractors in order to develop a complex system or solution. With the enterprise becoming increasingly integrated and interconnected, changes need both management and technical coordination.

While the language we use can be sensitive to the scope of the activity, the same change process has to be undertaken regardless of its scale. The word design is often reserved for the more significant technical decisions that have to be made, and we use terms like impact analysis or impact assessment for a change. Yet in both instances what is needed to implement a change is to determine how the requirements are to be satisfied, and that takes a design decision.

The term requirements allocation is used to describe the assignment of project requirements to the solution components. Following the analysis of the customer's requirements, high-level design decisions have to be made about the appropriate solution architecture to adopt. To ensure these design decisions are adhered to in subsequent project phases, requirements are allocated to components.

Requirements traceability then provides confirmation that the design features implemented can be attributed to the original requirements. Traceability provides confidence that the customer's requirements have been addressed and at the same time, ensures that there has not been any scope creep.

Figure 7.4 shows that the enterprise pattern can be viewed in terms of requirements allocation and traceability. The relationships between project and component can be seen to capture the high-level design decisions taken in specifying the solution a project needs to deliver. Each diagram takes a particular project's perspective even as all projects have to be implemented concurrently in the release. As a result, the separate design decisions made by the projects have to be compatible and support the argument made in Section 5.2 that separate solution architectures are less critical from an execution viewpoint than their consolidated and coherent release architecture.

The visual representation shown in Figure 7.4 can be seen to have an equivalent, more compact form in the release matrix shown in Figure 7.5. The latter figure highlights the particular case of project P2, where the requirements allocation is denoted by the presence of requirements assigned to components C2, C3, and C4, labeled as r_{22}, r_{23}, and r_{24}, respectively. These more detailed, derived requirements are a reinterpretation of the customer's requirements in terms of the solution architecture and in the language of the specific component. In the example shown, it is clear that the original (set of) customer's requirements for project P2 are distinct from the derived (set of) component requirements.

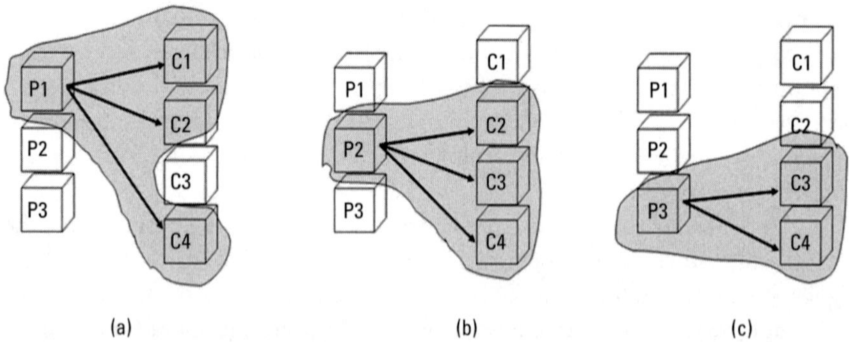

Figure 7.4 (a-c) Customer requirements perspective.

	Release scope	Priority	Comp. 1	Comp. 2	Comp. 3	Comp. 4
P 1	Project 1 requirements	2	r_{11}	r_{12}		r_{14}
P 2	Project 2 requirements	1		r_{22}	r_{23}	r_{24}
P 3	Project 3 feature set	1			r_{33}	r_{34}

Figure 7.5 Project 2 allocated requirements.

In many organizations, common requirements management practice sees the P2 project's customer requirements captured in a document, as would the derived requirements for each of C2, C3, and C4. It is in the detail of these documents that design decisions are usually captured and verified through traceability. Where requirements traceability is undertaken, often with the support of a tool to record the links, it is done at the level of individual requirements in these documents. The problem of this traditional approach is twofold. First, traceability is laborious task and the results difficult to maintain as requirements change. Second, the level of granularity at which traceability is performed can mean that key decisions can be buried deep within the documents requiring significant effort to review and understand.

Ultimately, requirements management best practices aim to ensure adherence to the customer's requirements and traceability to the solution or system architecture. To achieve that end unfortunately requires additional administrative effort to ensure design coverage of the requirements and provide auditablity of the design. In fact, the release matrix as presented in Figure 7.3 closely corresponds to a high-level representation of a requirements traceability matrix that is a common means of recording the traceability information. The key point is that instead of separately preparing such traceability data, the matrix notation suggests it may be possible to directly capture requirements and traceability together in this format. While this may not be suited for large-scale programs it can be a viable option in smaller programs, and this is discussed further in the next chapter.

The concise matrix representation also ensures that the documentation of requirements takes place within the framework of the architectural design decisions. The design decisions have to be made first, albeit at the architectural level, allocated to the respective components, with traceability subsequently providing a retrospective audit trail. The architectural decisions are made early, at the initiation stage of a project or change when, as part of the business case, suppliers need to be engaged to establish solution feasibility and estimate implementation costs. In contrast, traceability is conducted late in the delivery cycle, often only to ensure test coverage rather than design traceability, and being conducted at the requirements level can be far too granular to provide any management insight. Further, because these two activities are often separated in time, there can be reinvention and duplication of effort at each stage.

The disconnect between planning and execution discussed in Chapter 6, together with the fact that a project's solution architecture may be delivered incrementally across multiple release, can invalidate may design decisions made early in a project. The release matrix instead offers a means of capturing the high-level architectural design decisions made at the project level in its very structure (i.e., the cells which have content), while the elaboration of the customer requirements is the responsibility of component teams and is captured within each cell. The representation at this level is simply good requirements management practice presented differently. What is new in the release matrix is that in considering multiple projects delivering concurrently, the aggregated design decision taken across these projects, what we term the release architecture, can also be considered using the same notation.

7.1.2.2 Supplier Capacity Planning

The previous discussion had the customer's requirements as its focus and as such represents the demands placed on the component teams that support the enterprise architecture. The supply-side perspective of any agreement requires negotiation between parties in terms of the usual project management dimensions of scope, cost, and schedule.

Clearly the solution architecture has first to be concluded and the requirements allocation completed in order to determine the component teams to engage. This is often done as a precursor to the business case development by the sponsor of the initiative or in more centralized organizations, at the

PMO's instigation. The initiative has to be scoped and its feasibility ratified, with architectural design options investigated and reviewed before proceeding.

Once the preliminary solution architecture is determined (or at least considered worthy of investigating further), each impacted component team conducts further analysis to firm up the allocated requirements and come to a deeper understanding of what the customer needs so that the implementation can be scheduled and estimated within an agreed risk profile.

While the work entailed in developing a supplier's quotation calls for the participation of key technical and management resources, it is only when the customer (or the customer's proxy) determines to proceed with the solution that there can be any certainty in planning. While the customer continues to investigate options or has difficulty getting the business case approved, the supply team's plans are only interim and no firm commitment can be made to the delivery scope.

The primary goal guiding a supplier's negotiation team, variously led by business or relationship managers, is to satisfy a customer's demands—and where the supplier is external, win their business. To achieve that, the team must scope and architect an acceptable (component) solution that can be implemented within the customer's budget and schedule expectations. Irrespective of whether a supplier is internal or external, a profit center, or a cost center, there is a need to undertake capacity planning for both human and infrastructure resources in order to assess its ability to deliver the customer's requirements. Yet, in the period of uncertainty before customers agree to the design proposal and accompanying quotation (or internally, until budgets are approved), suppliers cannot commit to a firm delivery schedule.

As shown in Figure 7.6, the problem is exacerbated when a supplier has to deal with more than one customer. In breaking down the example we have been using for the enterprise pattern, it is clear that only the supplier of the C1 component has the luxury of negotiating with a single customer. For critical components, including core IT systems such as the customer care and billing system previously discussed when introducing the enterprise pattern, the

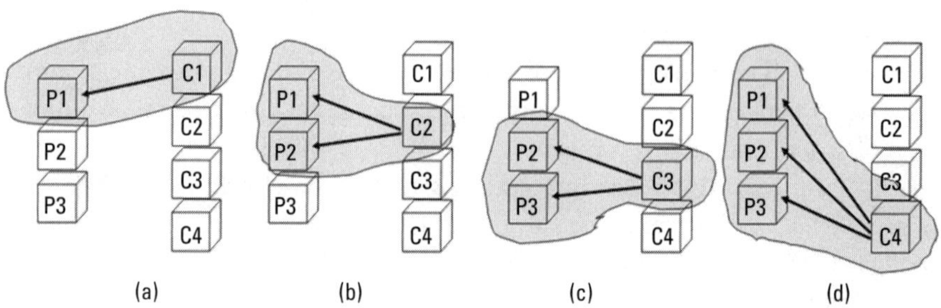

Figure 7.6 (a-d) supplier traceability perspective.

responsible supplier teams have to deal with the uncertainty of scoping and estimating several customer initiatives without any certainty of them getting the green light to proceed.

The catch-22 is that to get budgetary approval sponsors commonly require commitment to a delivery schedule. While each customer has the challenge of getting commitment from all its suppliers, these same suppliers have to balance any commitments they give against the demands of other customers. The approval of some initiatives but not others results in a continual cycle of replanning, as supplier teams need to renegotiate (or absorb) changes to schedules and budgets as a result of the changing (aggregated) requirements. When considered in this light, it becomes clear why change-driven, project organizations gravitate towards having resource pools so they can more flexibly manage their capacity to accommodate uncertain and variable demands. However, at some point there is a need to distinguish between capacity and capability, recognizing early experiences in systems development that highlighted the fallacy of simply scaling the workforce to deliver to schedule [2].

Figure 7.7 uses the release matrix notation to illustrate the multiple demands on the C2 component team. Here the challenge to satisfy its two customers becomes clearer when, importantly, it is realized that the combined or aggregated scope (r_{12} and r_{22}) has to be designed and implemented as a cohesive release within the agreed schedule. The benefit of all supplier releases to be synchronized can be understood from the (P1 and P2) projects' perspectives. The delivery of the C2 team's capabilities is simply and literally a component of the solution. To realize each of P1's and P2's solutions requires capabilities from the other component teams, which then have to be integrated and tested before the complete solution can be delivered to the customer.

While the scenarios described are by no means unique, and indeed are likely to be simplistic relative to the real-world planning and scheduling problems facing enterprises undertaking major transformation programs, the crux of the matter is that the release matrix notation offers a clearer representation of the opposing change (project) and architecture (component) perspectives. As

	Release scope	Priority	C1	C2	C3	C4
P1	Project 1 requirements	2	r_{11}	r_{12}		r_{14}
P2	Project 2 requirements	1		r_{22}	r_{23}	r_{24}
P3	Project 3 requirements	1			r_{33}	r_{34}

Figure 7.7 Component 2 supplier plan.

a notation, it crystallizes the critical issues facing the delivery of a change portfolio and embodies the abstract concept of an enterprise release. It is, however, simply a manifestation of the enterprise pattern, and where that is evident in an organization, there are myriad ways in which it can be applied, being simply a representation of a shared reality.

7.1.2.3 Representing the Enterprise Release

Considering the customer and supplier perspectives separately makes clear how intertwined the multiple relationships can become. We have previously related this to the dynamic system behaviors that an enterprise exhibits, but with the different parties having different goals and motivations, it begs the question of who cares about the aggregated whole. In other words, who owns the system? The common glue that binds the different parties together into one federated enterprise is their economic interdependencies. So, the simple answer is that it is the CEO and those delegated to oversee the strategic direction of the enterprise who should embrace the release matrix as it can represent the future trajectory of the organization as a whole.

The duality of the customer–supplier perspectives finds expression in Figure 7.8 showing the different objectives of the parties involved. Even where interdependencies constrain multiple initiatives and drive them to target the same release date, each of the parties participating in a release can be seen to have their own plan with its unique scope. The project teams look to deliver their specific solution release, integrating the capabilities provided by the separate components and operating seamlessly to deliver some predetermined business value. The component teams, while an instrument for the projects that need their capabilities, also have a responsibility for the quality and design integrity of their component. Each supplier focuses on their specific component release and provides the agreed capabilities required by the projects, but these only become operational once integrated and tested with other component releases.

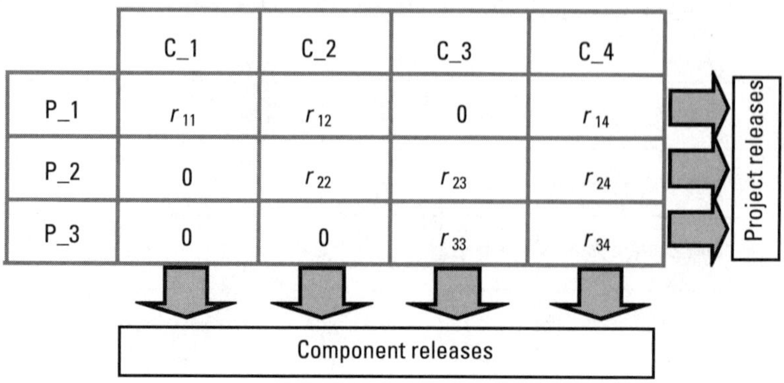

Figure 7.8 Project and component releases.

Any discussion of release dates clearly depends on the specific context. There is an ultimate scheduled date for the enterprise release; however, suppliers can be expected to have an earlier shared or synchronized release milestone. This is to allow the different component releases to be integrated and tested with each other. Prior to these formal dates, however, there may be informal release dates for particular features or requirements to allow project solution teams to gain access to critical capabilities and so reduce integration risk. All these separate and distinct release scenarios can be represented by the matrix notation while providing clarity and a shared understanding of the consequential impacts of any delays. Without a consolidated and holistic view of a release that includes the agreed, and possibly contracted, delivery dates it can be exceedingly difficult to understand the true status and risk associated with the enterprise release schedule.

So, as an aggregation of multiple project and component release milestones, the release matrix represents a master plan for the delivery of the enterprise release. By providing transparency of the state of the relationships and temporal dependencies between multiple projects and their suppliers, the notation offers new opportunities for balancing demand with capacity while improving collaboration within the enterprise. Using the notation to communicate the scope of a release, makes it clear to a project's sponsor and stakeholders that there are other initiatives which also have to be delivered in the same timeframe. While this is unlikely to dissuade the project manager from demanding that the project requirements are addressed, it can clarify the decision making that determines and explains why certain requirements cannot be met. The matrix makes clear that suppliers have to meet the expectations of multiple customers while also ensuring that the quality and design integrity of their own components are maintained.

The notation, however, makes no assumptions about how the individual releases are achieved, nor is it concerned with the development or management approaches used by the various suppliers. It is neutral as to whether agile methods are utilized or a plan-driven approach adopted for the component releases. These considerations are clearly the responsibility of each supplier who can make the determination based on local conditions and their weighing of time-to-market against architectural sustainability and maintainability. As such, the release matrix is more a management tool for planning and tracking the status of the different participants in the release rather than advocating a particular method or approach.

While the release matrix notation may not immediately offer benefits beyond clarification of the agreements, obligations, or contracts that bind the customers and suppliers together for a specific timeboxed release, it represents the stage on which the multiplayer game of enterprise transformation is performed. It provides the basis for simple new techniques that encourage a more integrated management approach. The notation is not arduous and it captures the different interactions and dynamics between those participating in the release. Different specialized disciplines can find their position or domain in the matrix, from the customer focus of project and requirements management to the quality and delivery concerns of suppliers responsible for solution components. The combination of the enterprise release as a delivery construct, and the release matrix as a notation that embodies the former somewhat abstract concept, provide a unifying concept for many common management approaches.

7.1.3 Managing with the Release Matrix

The holistic and explicit representation of the relationships between customers and suppliers captured in release matrix can foster a partnership model that approaches the spirit of communication and collaboration espoused by agile methods. At this level of abstraction, this representation has wider application than software development. While particularly relevant within enterprise boundaries, it can be more important where processes and services are outsourced to external agencies. Here the relationships become more formal and contractual, with a focus on delivery within the cost, schedule, and the quality envelope, while the manner in which the required activities are conducted become of less concern. The delivery focus purposefully treats suppliers as a black box, the emphasis being on the requirements being met rather than how they are met. This places the emphasis on delivery of capabilities and simplifies management by delegating responsibility for the supply of components—as a component based architecture intends.

In capturing the reliance that a release has on component, the release matrix has a correspondence with a manufacturing supply chain. There too the whole can only be realized through the integration of its parts. However, in the case of release management, there are distinguishing features that relate to the continual need to negotiate customer–supplier agreements that requires support and oversight. In the following section, we consider how the release matrix can be decomposed, and in keeping with the enterprise pattern, successively address the detailed layers of the solution architecture. The release matrix is shown to be a useful tool that can capture and clarify the volatile release planning cycle and help expedite management decision making. By allowing direct collaboration between the relevant parties, management overheads can be reduced and the scope optimized, ultimately improving the enterprise delivery capacity.

7.1.3.1 Enterprise Release Partnerships

Earlier discussion of release logistics highlighted the amount of preparatory and integration effort that needs to be undertaken to effect a release. As enterprise complexity scales, the problems faced in managing a single project are compounded as previously independent business solution have to be integrated at the enterprise level.

Each of the teams supporting these components can themselves have a reliance on subcomponents or modules and as such have to manage their own supplier relationships. This reflects the recursive nature of the architecture, and so the parties that are responsible for it be they internal teams, external service providers, or vendors. In software for example, the increased reliance on reused or commercially available software components or libraries means that the building blocks of any solution form a build of materials (BOMs) much as any manufactured product might. So, in managing delivery of the integrated solution, it is essential that the entire supply chain is mobilized, in change and enhancement as in new product development.

Figure 7.9 makes clear that decomposition of a single customer–supplier relationship is possible in both dimensions of the release matrix. While the supplier's architectural dimension can be decomposed to illustrate the BOM, so too can the change or project dimension be expanded to represent a customer's required capabilities. In the example, Project 2's scope is decomposed and separately allocated into the modules that comprise the detailed architecture for Component 2. Given the overloading of common words, terminology becomes constraining; however, the decomposition of the matrix allows relationships to be defined in ever-increasing detail, capturing the enterprise pattern's recursive nature, as illustrated by the nested Russian dolls shown in Figure 3.1.

A key benefit of the release matrix notation is its ability to adapt to the required level of abstraction being considered. Figure 7.9 makes it clear what is possible with tools that support zoom-in [+] and zoom-out [–] capabilities for columns and rows of the release matrix. A simple demonstration of the release matrix is available on the enterpriserelease.com website, where users can interact with prototype matrix tools that allow different management views of shared release information and delivery status. Clearly, such views can visually present status as a dashboard with traffic light reports that allow management to focus on particular problem areas. The benefit of this presentation is that it provides the wider enterprise release context, identifying delivery risk and their impacts, while also allows managers to drill down to the level at which the issue or problem is manifest.

A regular stream of timeboxed enterprise releases, sometimes called the release train, is analogous to manufacturing and the close partnerships developed between customer and suppliers. In attempting to reduce inefficiency and inventory costs, lean strategies such as just-in-time (JIT) [3] place great reliance on adherence to agreed schedules that can be disrupted by even small delays in the supply chain. Yet, with multiple customers to be satisfied, the enterprise pattern shows that it is closer to representing a supply network or marketplace than simply a single supply chain. Close partnering between customer and supplier

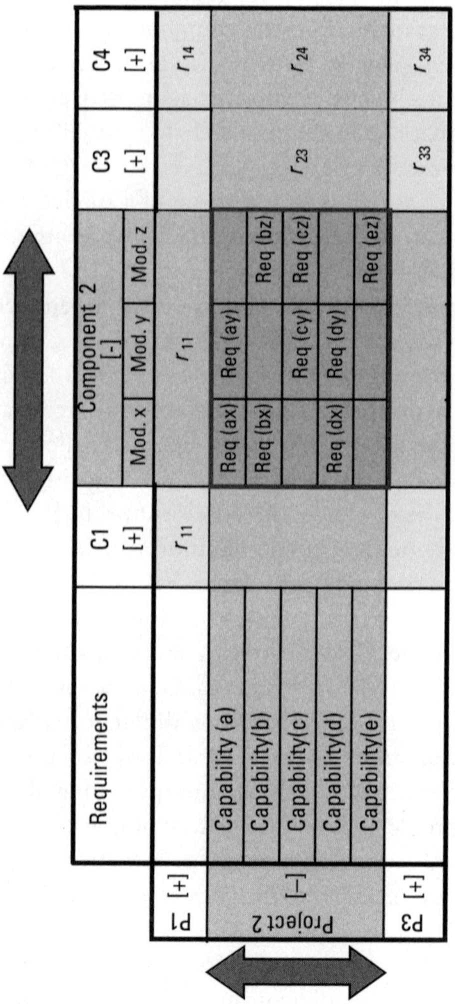

Figure 7.9 Expanded matrix showing capabilities and modules.

is essential in both, but multiple customers can place competing demands on each supplier that then have to be balanced and resolved as best as possible *and* delivered simultaneously. So while decomposition and the BOM give rise to a supply chain, the enterprise pattern is in effect multiple supply chains operating concurrently in the marketplace.

Unlike a supply chain where the relationship between parties can be expected to extend over a reasonable duration, the close partnering required for a release can be short-term. The participants in a release are only those customers and suppliers that committed to delivering capabilities in a release timebox. And as each release is a unique and separate proposition, the participants can change depending on which projects are involved and the solution components impacted. Further, enterprise releases are not dealing with prefabricated products, but with the shaping, execution, and delivery of the all-too-human activities and decisions, including:

- Determination of the strategic business requirements and their business value;
- Analyzing, estimating, and scoping alternative solution proposals
- Making design decisions that establish solution architecture along with development estimates and business change costs;
- Prioritizing business cases in order to maximize maximum value;
- Regularly reassessing and adjusting all initiatives in the portfolio to ensure they continue to align with strategic needs.

All that is to say that managing the logistics associated with an enterprise release can be significantly more dynamic and fluid than that of commodity-based supply chain. And with software-based IT solutions, the scope of the release can be reassessed midstream. Such adjustments to the release scope may be the result of changing priorities in the business but can equally be a result of component delays, emergent risks and issues, or design changes that arise in execution.

As a result, the need for customers and suppliers to work closely and collaboratively during the planning and integration of the solution is vital. The coupling between projects in the release portfolio can result in a domino effect and give rise to several rounds of planning which, as previously noted, can require more management and technical effort than the execution itself. These wasteful cycles of impact assessments and replanning only become more pronounced where outsourcing strategies make delivery reliant on external suppliers. Now the separate commercial interests mean that not all the parties are strategically aligned and the planning and estimating can be characterized by quotes, counterproposals, and escalations that can drive up the costs that such arrangements were meant to reduce—and delay the start of execution.

7.1.3.2 Expediting Decision Making

A truly democratic and inclusive release planning environment where all participants are involved can degenerate in what might be termed a strategic Scrum. A multidimensional version of what the Scrum methodology attempts to address in software development, it can be a free-for-all that results from the separate interests of the parties. In the enterprise environment where power is decentralized, such a scenario is not uncommon.

Strategic decision making in such an environment is clearly difficult. In the face of the diversity of views and goals, C-level executives and their leadership teams can have a tendency to withdraw to make decisions in isolation. Those charged with steering a strategic course face daunting challenges coping with the complexity and dynamism of the modern enterprise. There can be little time to understand and address operational issues, let alone assess and track the effect of a relentless stream of change programs. It is little wonder that the attention span of senior executives can be exceedingly short, as they need to rely on the advice of a bevy of managers and advisers with knowledge of their specific area of expertise. While this management model obviously works and is in play across most enterprises, it does not always support timely decision making. At least part of the incentive for lean or agile approaches is to break down management hierarchies. The frustration of software developers was that they believed customer requirements could be easily met if they were simply articulated and stripped of ceremony and management politics—development was faster than decision making. The hidden challenge in this viewpoint is that the remedy lies more in psychology and risk aversion rather than in improving the development process itself. Smart people can be quickly overwhelmed and struck dumb by the sheer enormity of data they have to contend with.

This is especially true in multiparty situations like the enterprise release, where far-reaching decisions have to be made based on few facts with often conflicting recommendations. So, when new strategies have to be implemented across departments or business units, there can be difficulty getting management to fully understand their implications, let alone behave in a cohesive manner that benefits the enterprise. The decision making process can take a long time given the briefings, reports, analyses, projections, and risk assessments that have to be prepared to guide, and sometimes protect, those making critical planning decisions.

The cost of slow decision making is difficult to measure but, depending on the corporate culture it can have a multiplier effect through the whole organization. Even where individuals may be decisive, there can be challenges from peers with competing agenda and interests that then require supporting information and assessments to address the opposing views. In multiparty planning scenarios the negotiations and compromises that need to be made and then approved can be debilitating.

The release matrix's succinct representation of the different agreements that are needed, and their implications and constraints, has been found to be a valuable decision-making tool. The aggregation of the different perspectives provides executives and their management with an information-rich dashboard that allows a focus on key issues while monitoring their impacts on related parties. But its greatest contribution might be as a tool to clarify and record agreements needed and then reached as a part of the complex multiparty negotiations that are part of planning an enterprise release.

7.1.3.3 The Release Planning Game

Release planning in the textbook pattern is as simple as it can get, with a single customer negotiating with the one supplier. Yet, as a testament to the psychology of unlimited wants that are a fundamental tenet of demand and supply economics, it can be challenging to achieve a realistic plan even in this simplest scenario.

A common planning and negotiation technique employed by agile software teams attempting to discipline the customer is to play the poker-chip game. This game is usually played by a development manager or team lead (the supplier) and the customer or sponsor's proxy (the project). The poker chips are used to represent the available resources or team capacity and the customer has to prioritize an often long list of previously estimated requirements by investing the available chips in developing selected requirements. This simple scenario serves to illustrate the challenge of managing the demand and supply equation and the different options available to the customer. They can challenge estimates by reducing risks or clarifying quality expectations, have the supplier team acquire and train new development resources, or prioritize their requirements and assigning the implementation of lower priority requirements to a future release.

Valuable as the poker-chip game is in highlighting limited supply and the negotiations that have to take place between the customer and supplier, it is restricted to a single customer–supplier relationship—it is textbook. Such a negotiation represents but a single cell in the release matrix, and an enterprise release needs all cells to be negotiated and agreed if it is to minimize execution risk. When we consider the totality of the participants involved in such a release, we essentially have a series of poker-chip games being played concurrently but where the outcomes can be interrelated.

The upshot of this is that the planning activities can result in a significant number of one-on-one meetings where project sponsors or their proxies attempt to reach an agreement separately with each supplier on the terms of the delivery.

There can be rounds of such meetings as any dependencies have to be addressed and as in the simple scenario described in Chapter 3 attests, the alternatives and implications of can quickly become a challenge. The usual manner dealing with the planning complexity is to largely ignore it and commence execution before for all parties can reach agreement across all dimensions of the release matrix. Instead, negotiations may be restricted to only the key supplier teams, with other supplier dependencies having to fall in line with the agreed schedule. Solution architectures do not identify impacts on smaller, downstream components of the solution, which adds risk to the release schedule. These omitted teams are expected to be able to scale their resources to meet the demands placed on them without consultation, which is not always possible—especially when the entire change portfolio is not always visible to them. It is no surprise therefore that stress levels increase as the release milestone approaches. And too often, the release is followed by customer disappointment with the capabilities delivered, the costs incurred, or delays in the schedule—and often, all of the above.

The game that "should" be played in planning an enterprise release centers round the relationships in the release matrix. Here individual agreements made between customer and supplier have to consider the wider context of the capacity of the suppliers to service multiple projects with a coherent and maintainable component release. This is a game that is more akin to the popular puzzle, Sudoku, where it is not sufficient to solve a problem in isolation of the broader context—winning in these scenarios entails winning at all levels.

The individual plans between the customers and suppliers that make up the cells in the release matrix are highly coupled. Based on each supplier's resource capacity, customers win when the projects they sponsor get the commitment of all their suppliers. Suppliers win when their commitments are achievable, utilize their resources, and their component architecture remains intact. Finally, and perhaps most importantly, the enterprise wins when the release is first defined to meet its strategic goals, and then the entire release has a reliable and low-risk plan to deliver clearly defined strategic outcomes.

We can envisage a new class of tools that support the release matrix notation and can track execution progress using the standard red, yellow, and green traffic light indicators. One scenario has individual squares in the matrix turning green as agreement has been reached and a customer wins as each requirement row displays green. Customers may find that their supplier(s) have committed their resources to other projects and as a result they cannot achieve a row of green cells. In such situations, the priority of the different projects or initiatives need to be assessed so that critical resources are assigned to priority areas. As the release moves into execution, each component team's progress can be tracked against customer expectations, with the cells now indicating delivery confidence. Again, where issues arise, the visibility the matrix provides can allow resources to be redeployed as necessary to priority requirements. The release matrix can therefore be seen to hold key management information about a release and further extensions of the notation will be discussed in the next chapter.

What the release matrix variant of the planning game makes clear is that not all players can be winners or even get to complete the game. Even with the benefit of this new tool for negotiating multiparty arrangements, the management overheads in achieving a completely defined and locked-down enterprise release plan may simply be too great. Instead, a risk-based approach to planning is needed, and once a suitable level of understanding is reached, execution can commence and the release matrix can then be used in an agile manner as a tracking tool—albeit without a firm set of agreements in place at the outset.

From an enterprise perspective, the release planning game is in fact serious business. The winners should be those strategic initiatives, not simply the projects that have good negotiators or managers who can bully their way to securing agreements. The rules of the game need to be adjusted to ensure such initiatives get priority over those with lower business value. Further, after all the rounds of negotiation that take place in the planning game, there need to be ways to ensure that the end result represents an optimal and strategic utilization of enterprises resources. For example, did those projects that could not get all suppliers to commit to delivering their requirements in this release—that is, achieve a full row of green lights—release the resources they secured from suppliers who did commit to implementing their requirements? Or is it possible those suppliers simply "wasted" those resources and now have unplanned spare capacity? The volatility of planning can make it difficult to keep track of the proposals received and estimates given over several planning cycles, and the matrix notation can offer some support in keeping track of just what is in or out of scope of a release.

7.1.3.4 Optimizing Release Capacity

In much of planning, the talk is about maximizing the utilization of resources—we do not often talk about optimizing the features that we deliver. The scenarios discussed above point to the increasing management overheads that result when attempting to plan in the enterprise. In the confusion of multiple rounds of scoping, designing, and estimating proposals and requirements from projects, the component teams who work across concurrent projects can find it difficult to maintain a baseline of the finally agreed scope of a release. Then, the added difficulty of negotiating around a matrix of relationships can result in the need to make unilateral, simplifying assumptions.

More often than not, customer–supplier negotiations are conducted independently of, and isolated from each other, even as the interdependencies have consequential impacts on other agreements. The release matrix can therefore provide a means to consolidate multiparty plans and provide overall status of release planning. The matrix provides some self-organizing features where consequential impacts can be identified, affording a measure of "peripheral vision." The different parties are at least able to recognize the knock-on effect of changes to schedules and where, again as in the scenario in Chapter 3, the inability of one supplier to commit to a requirement results in related requirements having to be removed from other supplier's release scope.

The game that is enterprise release planning can be a highly interactive, multiplayer game. And the release matrix simply provides the board upon which the players must necessarily interact—even if they cannot play the game well, or even know its rules. To play well requires new techniques and tools for navigating the release matrix that are the subject of further research and investigation—and readers are invited to join the discussion at enterpriserelease.com.

Optimizing the resulting enterprise release plan requires careful balancing of the competing forces of change and alignment with those of complexity and stability, and it is no trivial task. Component teams cannot simply accommodate allocated requirements without considering their specific component architecture. Yet developing an optimal enterprise release plan needs to be achieved in an expedient manner. The ultimate goal of the planning game in the enterprise is profitability and growth, not a perfect plan.

Offering a shared, representation of the state of the planning, the matrix notation can support rounds of planning that gradually establish an agreed enterprise release plan. It provides a framework for more agile or risk-based planning, defining the scope of the release including the requirements that each component team is to deliver. Each enterprise release plan targets a release timebox on the enterprise roadmap. All agreements made between customers and suppliers are therefore futures contracts, commitments to deliver the necessary capabilities at the agreed times. However, the real value of the notation is that a planning cycle can be rerun as circumstances change or issues are raised that affect the delivery dates.

From a customer standpoint, the sizing estimates provided by suppliers cannot guarantee that the needed resources and infrastructure will be available at some future point in time—quotes always have a validity period. As shown in Figure 7.10, the planning phase of the release creates a dynamic, multidimensional demand and supply problem. The estimates provided by suppliers for allocated requirements have to be agreed to as a "set" as projects aim to ensure a synchronized and integration delivery. At the same time, the total demands placed on the component supplier by the solution architecture of each project have to be within the team's ability to scale its resources. The interactions between these two dimensions results in the difficulty achieving a stable plan. Aside from resource capacity, there is also scheduling to be taken

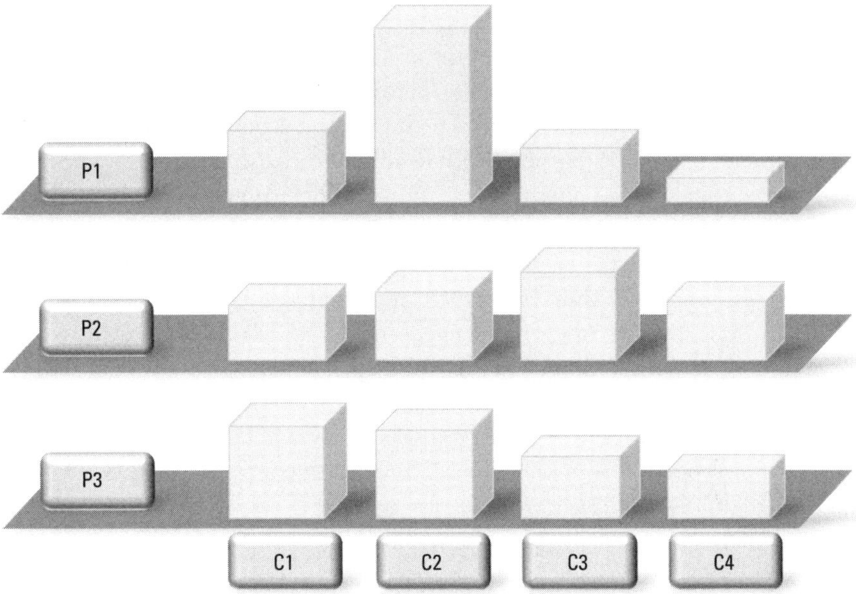

Figure 7.10 Managing multidimension supply and demand.

into consideration. The problems planning and coordinating just two supplier teams was discussed in Section 3.1.2. These problems compound as the number of participants rise. In Figure 7.10, the high demands that project P1 places on the supplier of the C2 component may make it impossible for other requirements allocated to the C2 team considered for that release. Unless an alternate, partial solution can be designed that reduces the demands on the C2 team, it may be expedient to withdraw the requirements allocated to the other component teams, giving rise to another round of planning as new requirements now take they place.

Like the original planning cycle, descoping can be difficult as the removal of all allocated requirements relating to a high-level customer requirement can be difficult to trace. Yet it is a relatively easy task when the release matrix is used: first the component team that has the issue identifies those allocated requirements that are problematic, follows along each row to find the originating customer requirements, then goes back from that point to discover if there were any other requirements allocated to different teams. By maintaining the allocation relationship, this is relatively easy in the matrix notation, but in document-centric requirements, the requirements that need to be removed can be deep in a requirements specification, making it likely that some redundant changes are implemented. This wasteful scenario happens simply because it can be too administratively difficult to identify and remove redundant requirements.

7.1.3.5 Agile Release Execution

The planning of a strategic release has similarities to many of the issues individuals face in task prioritization and personal time management. Clearly, it requires strong discipline and practice before it can become a routine. However, a relentless focus on priority tasks with deadlines can also become mentally exhausting, and in such an environment some self-determination and autonomy can provide the means to improve morale and productivity.

A common time management technique is the big rocks prioritization [4] that identifies the most important goals, represented by big rocks, and prioritizes them first, followed by progressively smaller and less important tasks. In a compelling demonstration of the big rocks principle, a glass cylinder (whose capacity represents time as a limited resource similar to the poker chips previously discussed) is filled first by large rocks, then smaller stones, then gravel, which is followed by fine grains of sand. Finally, even as the container is visibly full, water is added to drive home the message that there remains spare capacity when we plan our time in accordance with priority.

Big rocks prioritization is effectively what is done as a part of release planning. Figure 7.11 illustrates two rounds of release planning, applying the big rocks principle with strategic initiatives being accepted by a team, with the cylinder here representing the resources that team has available. The activities that need to be planned and executed are represented as spheres with the size representing the resourcing estimated to be needed—the big rocks analogy is weakened somewhat because priorities here do not simply relate to size. Nevertheless, the first cylinder shows the results of an initial round of planning where strategic requirements are committed to the release. The second cylinder shows the subsequent point at which the team appears to have reached its capacity as lower priority requirements are added. However, there remains some capacity

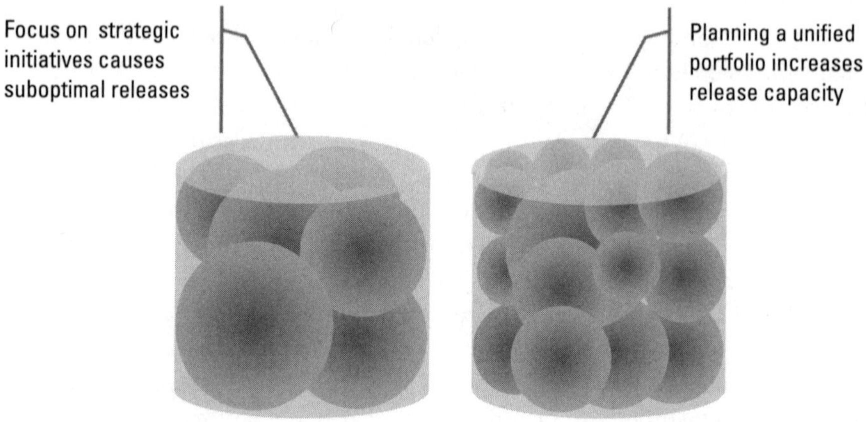

Figure 7.11 Optimizing release capacity.

within the team that arguably is not worth the planning effort required to assign fine-grained tasks. Better to leave that to individuals who can be given the discretion to work on selected changes they can pick off this backlog.

The divergence between release planning and the big rock approach is that those doing release prioritization do not often have access to the finer grained activities. While these smaller changes need to be performed, most organizations manage them separately from the projects executing the more strategic, or at least more costly, initiatives. In the big rocks analogy that would mean we would arrange multiple graduated cylinders representing the different teams (including for example, resources in project, component development, maintenance, or customer support teams), but they each only have access to one size rock that has been assigned to them. This separation means that the teams do not have the flexibility to optimally use their resources—those times when an activity may be blocking them as a team or individually. Demanding the focus be on the work that management assigns can be less optimal than allowing some limited choice to utilize spare capacity—much as we recognize that need in the personal domain.

The unified change portfolio that was advocated in Chapter 5 essentially aims to consolidate the different smaller change processes from across the organization in one change process. This philosophy is closely aligned with the personal time management system described by David Allen in *Getting Things Done.* While the scale of the changes may differ, the fundamental principle applies: establish a unified, prioritized list of actions that can then undergo a consistent prioritization and initiation processes. This can help ensure that there is no wasted capacity that would otherwise be absorbed by the schedule

There is a deeper significance to having a unified portfolio and the ability for teams to have some, limited choice over the requirements to be addressed. That insight lies in the recognition that all requirements can and are decomposed into smaller grained activities. So that in projects, the big rocks eventually become smaller rocks, too. And therein lies the potential for the more agile execution of a release—break the original and somewhat arbitrary project scope into smaller component tasks that can then be more opportunistically executed by the component teams. This does not mean the dissolution of the project concept, which remains relevant as a business construct and manages the ultimate business outcome. Instead it relies on the role of release planning and design to determine when capabilities are actually delivered and allows the release scope to be optimized.

This may appear somewhat revolutionary given the presiding project paradigm, but it is an activity that takes place routinely in fine-grained change management. There it is the release package that is the delivery vehicle, albeit that these changes can often be independent; it still provides a working example to contemplate. In the enterprise, the need to track the relationship of the smaller changes to their parent remains, providing clarity as to when a higher-order requirement is met; the matrix notation can accommodate that adequately. But what the time management can teach us with the likes of the big rocks principle and unified task lists is that so long as there is a solution architecture and a coordinating mechanism such as that provided by the release matrix, then team productivity may be best achieved when they are allowed to organize themselves and determine the best order to execute their tasks. This hybrid of a centrally coordinated task list provides more agile teams with some flexibility to work on a prioritized list of fine-grained activities removes project from execution completely. While they capture customer requirements and desired outcomes, their execution can be assigned to component teams which optimize their designs and delivery capacity, delivering into predetermined and experienced release teams. This then is the ultimate vision of ERM as an agile enterprise life cycle focused on delivering stable, low-risk releases that meet project demands without being dictated to as to how best to deliver them.

References

[1] Reingruber, M. C., and W. W. Gregory, *The Data Modeling Handbook: A Best-Practice Approach to Building Quality Data Models,* New York: John Wiley & Sons, 1994.

[2] Brooks, F. P., *The Mythical Man Month,* Reading, MA: Addison-Wesley, 1995.

[3] Monden, Y., *Toyota Production System: An Integrated Approach to Just-In-Time,* 3rd ed.,

[4] Institute of Industrial Engineers, 1998. Covey, S. A., R. Merrill, and R. R. Merril, *First Things First: To Live, to Love, to Learn, to Leave a Legacy,* New York: Simon and Schuster, 1994.

8

The Enterprise Release Life Cycle

> *The improvement of understanding is for two ends:*
> *first, our own increase of knowledge;*
> *secondly, to enable us to deliver that knowledge to others.*
> —John Locke, philosopher

The project or development life cycle has been touched on in the early chapters. Indeed, it is impossible to discuss the delivery of strategic change without considering how strategy is translated into action and ultimately business value. The different stages, from the formulation of a strategy through to its planning, execution, and delivery, have to be viewed as a continuum if we are to successfully deliver strategic change.

However, ERM is founded on the notion that it is the release that ultimately delivers change in the enterprise and needs to guide enterprise evolution. The enterprise pattern suggests that the project is no longer the appropriate paradigm to deliver enterprise change. Singularly, projects in a portfolio can be too granular and even a program has to be viewed in light of concurrent initiatives—if only to ensure unexpected dependencies do interfere with its plans. It is therefore the release to which we must turn to understand and visualize enterprise evolution.

A release-centric perspective raises the abstraction level and relates strategy to the enterprise roadmap—the planned stages in an enterprise's evolution. Controlling the changes that are to be ultimately delivered in each release ensures that strategic priorities are evident throughout the release lifecycle. Much as a project takes a higher-level abstraction of the details of the development phases (to address organizational change, for example), so the enterprise release lifecycle must bracket the different projects and change initiatives targeting that release—making clear that the individual initiatives are subordinate to the collective release.

Just as the complexity associated with delivering a change portfolio can be difficult to manage, so its multidimensional nature is difficult to represent and visualize. The previously introduced release matrix notation is valuable when extended in the temporal dimension to describe the different stages in the release life cycle. Capturing the essential relationships and interdependencies between the parties participating in an enterprise release, the result will be shown to be a multistreamed life cycle describing the planning, execution, and delivery of a change portfolio.

By capturing the enterprise pattern, the enterprise release life cycle brings together the different roles involved and offers an elegant visualization of the different perspectives that make for strategic delivery in a release context. This aims to redress the fact that separate disciplines have been allowed to take their own unique slant on managing what should be the one thing—strategic delivery of change. The project paradigm has resulted in a collection of project-centric practices, addressing the full range of change granularity from portfolio, program, and project. Separately, architectural design-centric practices have operated within what might be termed a system paradigm that focuses on maintaining design integrity and continuity. And even where each is applied diligently, they rarely operate effectively together, failing to achieve a single, shared view of the change life cycle. Indeed, it can seem that the more diligently each of these disciplines is applied, the further we get from an understanding of precisely how the current state will be transformed into the target state. Where there once might have been the common touchstone of a systems engineering process that has been lost as the different disciplines have diverged over time. Certainly, in the throes of execution, the artifacts different flavored best practices demands, such as plans and designs, can become disassociated from the reality on the ground. While the value of each discipline is not questioned, the manner in which they are now implemented has become disjointed and detached from reality. As each continues to mature (apart) they can become too demanding of practitioners to apply in the midst of real-world execution problems—making it too easy to discard them completely as idealized or theoretical constructs.

In this penultimate chapter, we attempt to consolidate the concepts previously discussed and provide the framework for managing the enterprise release. The objective is to lay the foundations for a new class of life cycle models that addresses the enterprise pattern. To better understand the need, traditional development life cycles are first reviewed in terms of the delivery patterns they support. The enterprise release life cycle is then elaborated and shown to reframe many familiar project activities while addressing the multiplicity inherent in the enterprise pattern. This life cycle must be multistreamed in order to encompass the traditional development with critical release milestones to which project schedules need to align.

Finally, the enterprise roadmap that defines the schedule for future enterprise releases is related to strategic planning and is shown to provide an overarching framework for managing change and transformation.

8.1 Delivery Life Cycles: Past and Present

The chaotic manner in which projects are executed in the enterprise, combined with the poor success rates reported provides evidence that there is a problem. The remedy that we employ, however, depends on whether one considers the problem to be a result of not applying current best practices effectively or whether these practices are in fact inadequate. It should be clear at this point in the argument that this book takes the latter view.

The heresy at the heart of the argument results from frustrations with the difficulty organizations face in adopting current management best practices, and how quickly they are discarded by practitioners. Surely the right way of doing something cannot be that difficult to implement? The promised benefits should not be so difficult to attain.

The alternative view that this evidence suggests is that current delivery life cycle models may not be adequate. Proponents of an agile or lean management philosophy would argue that traditional management practices require too great an overhead for the benefits they offer. But that may in fact be too generous—current practices may be the right solution applied to the wrong problem.

The challenge is to identify processes and develop models that are at once integral to the task at hand, and yet provide management information, negating the need for further interrogation of the status. The trouble is that it is difficult to test them for validity in a rigorous manner given they are human processes and there are so many variables. This explains why there is a prevalence of "flavor of the month" management approaches and models that introduce new buzz words but usually little in the way of new insights.

8.1.1 Traditional Life Cycle Models

The system or software development life cycle (SDLC) has been an important touchstone for the different disciplines that must work together to deliver a business solution. They allow these specialist roles to coordinate their activities, defining the hand-off between them so that their separate contributions can ultimately contribute to the quality of the solution.

Life cycle models represent the distillation of hard-won experience. By applying them, we aim to leverage this knowledge in our own projects. The number of alternative life cycle models that have been proposed gives the sense that they offer alternatives for a range of conditions and environment. Each promises a formula for project success, and these models run the gamut from rigorous and detailed so-called methodologies (more accurately called methods or processes) that once required bookshelves to the simple tenets that gave rise to the agile methods.

What makes life cycle models interesting is that in the absence of any real test of their validity, their adoption can often depend on their appeal to our personal philosophy, or how specific project roles consider things should work. Naturally, each role in the life cycle has its own perspective and reason why their work is critical to the success of the overall project. There are myriad ways in which the core development activities can be defined and success itself can be highly subjective. The methodological debates once arguing the merits of functional decomposition versus object-oriented design are reprised today in the debate over the merits of plan-driven versus agile methods. Life cycle models can attract adherents with near-religious fervor and the selection of a life cycle method to apply in a project or as a standard across the organization can be a difficult decision. A popular joke among those of us who have been working this space goes like this:

Q: What is the difference between a methodologist and a terrorist?
A: You can negotiate with a terrorist.

However, in reviewing the different life cycles we will not be focusing on their relative merits but rather their form, or structure. In the broader context of the enterprise release, "how" a team undertakes the development or delivery of a solution component is of less interest than the manner in which these components come together as an integrated release. Instead, we need to probe the underlying structure of the different models and their relation to the delivery patterns described in Chapter 6. We attempt to take an x-ray view of the alternate model, that looks past the processes, deliverables, and milestones that give it flesh, to understand what problem our traditional life cycle models are attempting to address and how that contrasts with the generalized life cycle proposed.

8.1.1.1 Textbook Development Models

The waterfall model [1] illustrated in Figure 8.1(a) is clearly the simplest textbook life cycle. Few would argue with the basic premise at the heart of this well known but much-maligned model. It "is" the textbook pattern, addressing the needs of a single customer with a single solution. Taking logical problem-solving steps to the customer's requirements, it defines the phases of systems development in a sensible sequence. Projects start with the customer's requirements, the problem space is analyzed, and a proposed solution is then designed

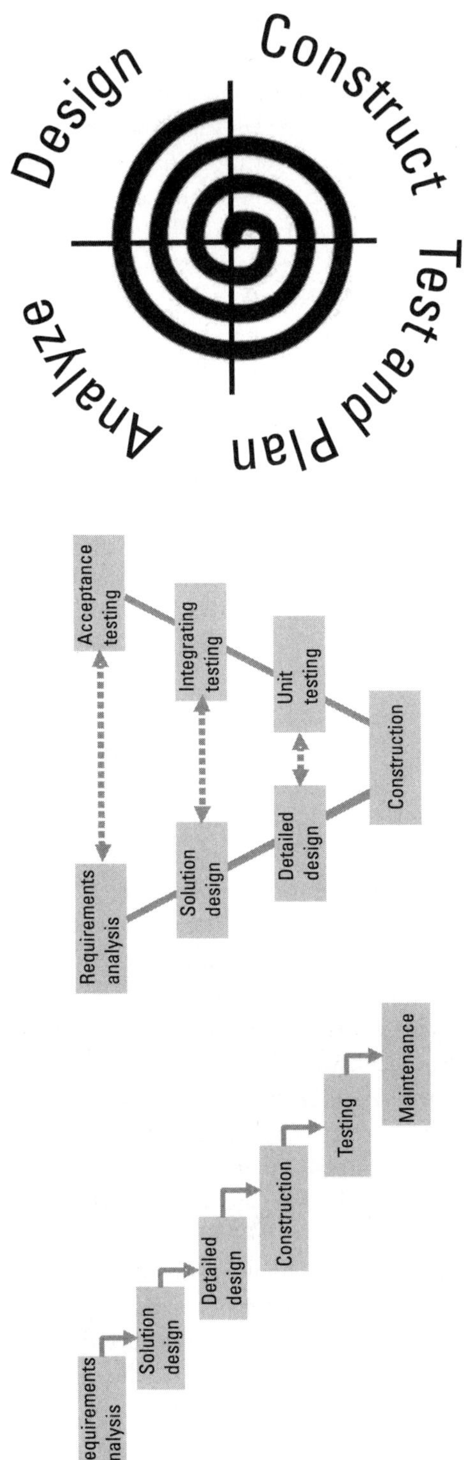

Figure 8.1 (a–c) Different life cycle representations.

before it is constructed, tested, and deployed. Difficulties only start to arise when we scale this sensible sequencing of development activities and require an entire development team to undertake the same activity in unison. Yet, for all its critics (including its originator), the waterfall model has become the classic life cycle, establishing the foundation for other more elaborate models.

There are several variants of the waterfall model and two popular representations, the v-model [2] and the spiral model [3], are also shown in Figure 8.1 (b) and (c), respectively. Each makes a key contribution to better understanding key aspects of the development life cycle, which is reinforced visually. The v-model focuses on the relationship between the requirements at different levels of elaboration and their validation at corresponding testing phases. Attempting to ensure that the participants in the life cycle form a cohesive team and understand the need to link requirements to the capabilities delivered, the v-model relies on traceability across the life cycle to ensure quality. The phases remain largely the same and are performed in the same order. So, apart from reinforcing the need for different roles to focus on customer requirements through the life cycle, it is not significantly different from the waterfall model.

The spiral model, on the other hand, can be credited for being the inspiration for many of the concepts in agile methods. It is, more accurately, a life cycle, graphically representing the iterative and incremental cycles that repeat the phases of the waterfall life cycle ad infinitum. Its cyclic nature recognizes that development can be an ongoing elaboration of an evolving solution, as opposed to a number of finite phases that end at project completion. The spiral model allows the development to seamlessly transition from an initial prototype through to an operational solution and continue on to the different maintenance and enhancement releases. It is an important step in identifying the commonality between greenfield development and ongoing maintenance activities. As development projects increasingly give way to the implementation of packaged solutions in the enterprise, the distinction between development and change fades. As a result, the spiral model remains relevant today, albeit in new incarnations such as Scrum, where each sprint can be viewed as a cycle of the spiral.

What becomes clear in reviewing these seemingly different life cycle models is that they are in fact surprisingly similar. The waterfall model provides a convenient building block, or template, that is rearranged, bent, or curved to reinforce a particular perspective. The apparently distinct models have essential, topological similarities. Importantly, they all support the project paradigm, sharing the fundamental premise of a single customer (be that an individual or organization). Even where multiple projects may each conduct separate life cycles concurrently, each independently addresses a specific customer's needs. In other words, these models are attempting to solve the same problem in different ways, making them of the same class of solution that aims toward the textbook pattern.

From this structural perspective, even agile methods are of the same class and remain aligned to the waterfall model that they deplore. Whilst loosening up the criteria for the phase gates and undertaking development at finer granularity, most agile methods fundamentally follow the same logical sequence of activities. In embracing change, agility is even more aligned to the textbook pattern—considering the customer relationship as critical to prioritize and guide different iterations.

Recognizing the structural similarity between current life cycle models does not diminish the contribution they make. Each of these models has had a profound influence on the way we understand development and, like the textbook pattern, remains pertinent to many projects conducted in the enterprise today.

But while we focus on different ways to deliver a solution, the nature of that solution has changed in a fundamental way in the last few decades. We may have let the "how" of projects distract us from the "what" that needs to be delivered (i.e., let the process debates lose sight of what has to be delivered). To better manage change and transformation, we need to go beyond the textbook pattern and acknowledge the systemic properties of the modern enterprise. With today's familiar life cycle models supporting the project paradigm, finding models that address the enterprise pattern requires that we look to the future.

8.1.1.2 System Integration Models

Ironically, the chronology for life cycle models did not follow the expected evolution and start with the simple textbook pattern. That delivery pattern did not become common until computer systems became more widespread and businesses initiated relatively simple projects to exploit the new data-processing capabilities. Much earlier, in the Cold War era and even before the waterfall model was given its name, engineers were undertaking large federal programs developing complex systems for the defense and aerospace industry. These large multidisciplinary programs involved the integration of hardware and software components often developed by separate contractors.

In many ways this was a golden age of systems engineering, a time when the integration pattern had to be addressed and there was a general confidence that the industry knew how large programs should be managed. While systems of that era were technically complex, carefully selected and experienced prime contractors were made responsible for system delivery. In these halcyon times, before software had made any significant impact, the system perspective provided a shared bond between technical and management disciplines, spanning what was known as the "acquisition life cycle."

The large-scale programs undertaken by the U.S. Department of Defense (DoD) required oversight and coordination, prompting the codification of accepted best practices. These best practices have remarkably survived the test of time and, with some minor change to language such as replacing system integrator for prime contractor and outsourced for contracted, we find that the modern enterprise has now also arrived, and as we will see, surpassed the complexity of the integration pattern.

Figure 8.2 shows the system development life cycle known as DoD-STD-2167A [4] that serves as an example of an integration model, since incorporated into ISO/IEC 12207 [5]. The model can be seen to retain key aspects of the waterfall life cycle with the key differentiation that it visibly sets out to address the integration pattern. Separate branches are shown for component development that are wrapped or bracketed by system-level activities. The system life cycle commences with system and specification that defines the system context before multiple component development streams fan out. Each branch could comprise multiple, separately contracted and developed components, denoted as HWCIs and SWCIs for hardware and software configuration items, respectively. And while the terminology may be unfamiliar today, what is clear is that the different component development streams are undertaken concurrently with regular reviews, ensuring that their separate life cycles were synchronized and aligned at key release milestones.

While there are echoes of the waterfall in the DoD-STD-2167A example, there is also something unique in its explicit recognition of a componentized solution and the resulting need for integration. The system perspective is clearly present at both the start and conclusion of the life cycle. Here the waterfall is in fact recursively applied for each component rather than iterative, reflecting the self-similar nature of the enterprise pattern, illustrated in Figure 3.1 as the Russian dolls. Further, the need to synchronize the lower-order life cycles to ensure coordinated delivery corresponds to the need in the enterprise to align independent project schedules with timeboxed enterprise releases. This integration model can therefore be seen to be distinct from the textbook life cycle modes discussed previously and representative of a class of integration models.

As onerous and rigid as the DoD standards were, they offered a comprehensive and coherent model for system development and integration. The contractual framework ensured that the role played by the different disciplines across the life cycle was made clear, and the technical complexity could be kept at bay by appointing a proxy, a prime contractor, or system integrator, that took on the system engineering and program management responsibilities. Such delegation is only possible in the integration pattern because the customer remains a single point of control.

A system integration life cycle is fundamentally different from the traditional models considered earlier. Rather than being iterative, it is in fact recursive in nature, introducing new system-level activities to first decompose and then integrate complex solution architectures. These activities bookend multiple component-level life cycles and as such represent a new class of life cycle

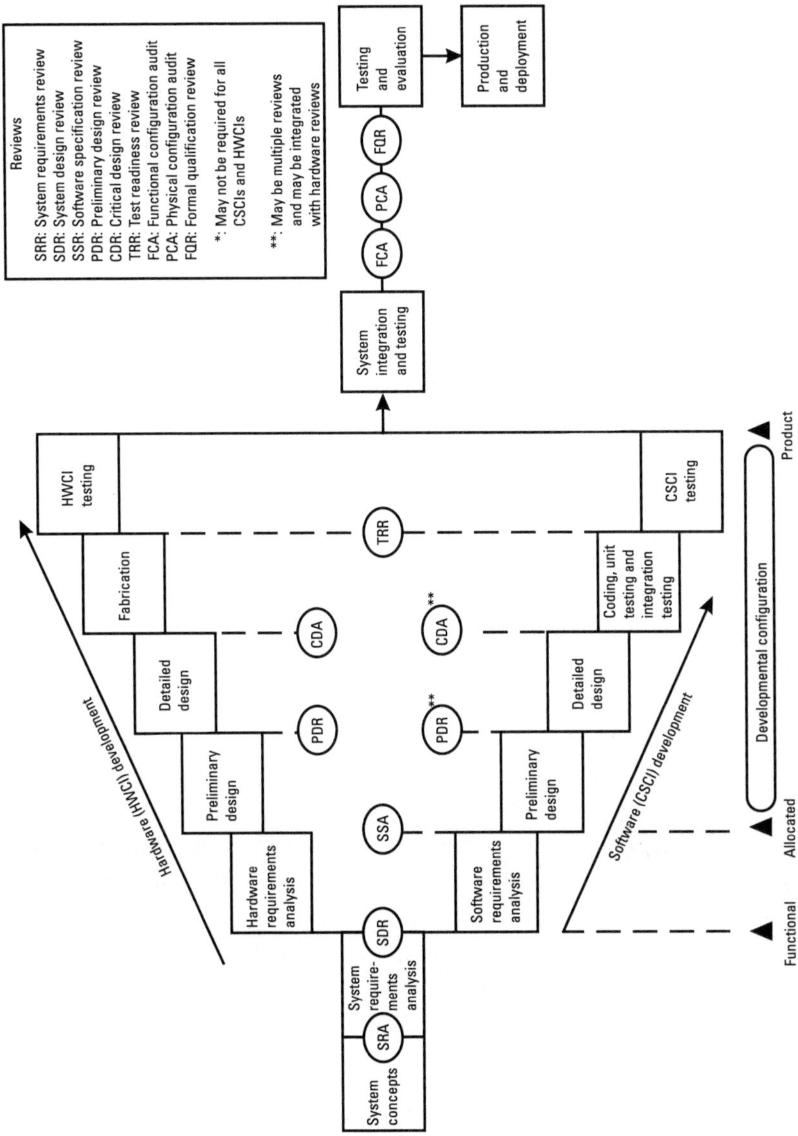

Figure 8.2 System development life cycle DoD-STD-2167A.

models that nonetheless retains the customer as a central controlling agent, able to step in as needed to assume command and ensure the integrity of the system.

For all the system complexity that models like DoD-STD-2167A can manage, the customer is able to be shielded from the reality by the delegation of the technical complexity to a proxy. That will be seen to be more difficult in the modern enterprise with its decentralized and distributed decision making across business units. Under such conditions, multiple business sponsors are drawn into this complexity as their requirements and priorities are challenged and negotiations require their direct attention.

8.2 The Enterprise Life Cycle

The problem with having customers shielded from complexity is that they do not have to acknowledge its existence. In the case of the DoD standards, the systems built for defense and aerospace were understood to be complex from the outset and so the mechanism necessary to cope with that complexity were then established. In the business world, where the textbook pattern has prevailed, complexity has been building up slowly so that, like the proverbial boiled frog that does not recognize the slow rise in temperature that will eventually be its demise, organizations are simply hoping complexity will go away—perhaps with the next upgrade of their new enterprise system.

However, there are several key distinctions between the development life cycles of the past and the new class of enterprise life cycles needed going forward. One such distinction is that few solutions are developed in the enterprise anymore, and even when there are development projects they are often more correctly efforts of configuration and integration. While the IT industry has matured to the point where commoditized packaged solutions exist for most business needs, we have been slow to update our management paradigms. As evident in the enterprise pattern, projects today have a closer similarity to a change process than they do to new development. And so not surprisingly, we find a greater need to apply release management principles to avoid the (real) enterprise system from fracturing and losing its design integrity.

Another point of differentiation between the next-generation, enterprise life cycle model that we need from those previously considered is that the latter make unrealistic assumptions about management control that are rarely accurate in the enterprise context. Irrespective of their arrangement of the waterfall template, both textbook and integration models take the perspective of a singular customer who has old-fashioned command and control over an entire project or program. The presumption of control has encouraged the project paradigm, and it can be argued that recent agile methods have gone furthest in that direction, and in seeking simplicity have retreated to the textbook pattern.

As captured in the enterprise pattern, modern business organizations are highly dynamic with power and control decentralized. Management proxies, like prime contractors and system integrators, are less effective as customers are routinely called upon to make difficult prioritization decisions that cannot simply be delegated. We have gone from technical complexity to management complexity, and business leaders will find it increasingly difficult to shield themselves from the problems of execution and delivery. The need for difficult prioritization and trade-off decisions will make their way into the executive suite as initiatives compete for resources and a release window.

Any enterprise life cycle must embrace collaboration, which becomes essential as the number of decision makers increases. The enterprise pattern has resulted in the need to manage dependencies through release planning and design, all of which will require negotiation and prioritization between the different customers or sponsors needing access to shared enterprise resources. An active dialog between the different parties, both customers and suppliers, is likely to extend over the entire life cycle, which, from previous experience, is going to be some new arrangement of the versatile waterfall life cycle.

In looking to animate the enterprise pattern and establish the life cycle stages through which an enterprise change portfolio must transition, we need more than a system development life cycle. The multiplicity inherent in the pattern calls for something more akin to a life cycle model for a system of systems. Yet, at the current level of maturity of release management, they would not be able to adopt any life cycle model that was overly complex. As the release matrix has been shown to provide a means of succinctly capturing the change portfolio and describing key features of the enterprise release, the matrix notation will again be exploited and become the basis of the proposed enterprise release life cycle termed the matrix life cycle model.

8.2.1 Animating the Enterprise Pattern

Models by definition offer a simplified representation of reality. They are aimed at highlighting key elements, but remain silent on those aspects that are not under consideration. Capturing the essence of a problem is the goal of a good model and explains the enduring nature of traditional models like the waterfall, spiral, and v-model. It also explains why they have remained silent on the issues of multiplicity and dependencies that we are now interested in exploring—that was not the problem space these earlier models set out to understand.

Viewing the enterprise holistically and treating projects as if they were changes to the enterprise architecture has resulted in new delivery challenges captured by the enterprise pattern. The management goal now is to develop a model describing how the enterprise as a whole can evolve strategically and find a representation for what is the enterprise life cycle. Such a life cycle can better link strategy with delivery, holding the promise of connecting the dots between portfolio planning and execution.

To explore the problem, we start with the enterprise pattern that captures the dependencies and interactions between projects in the portfolio. This needs to be understood from a temporal perspective, animating the pattern to determine how it plays out over time. The delivery of a portfolio of separate change projects can be expected to manifest as multiple, interdependent life cycles—more like a system integration model than a traditional development model. And like the generalization that gave rise to the release matrix, such a life cycle will demand a more nuanced vocabulary to describe the dynamics and interactions between the parties.

It should be made clear that in starting with the enterprise pattern we are tackling the problem of an interdependent portfolio. While the parties involved in the pattern may have acted autonomously in the past, their shared reliance on core systems and services bound together the portfolio with dependencies and prompted the need for an enterprise release. This demands a unique life cycle model rather than simply conducting a collection of concurrent independent projects using multiple instances of a more traditional life cycle. The latter is a legitimate example of portfolio execution, if it does not suffer the dependencies apparent in the enterprise pattern.

In what follows, we establish the foundations for an enterprise release life cycle that describes the definition, planning, execution, and delivery of an interdependent enterprise change portfolio. The implications of a multistreamed and intertwined portfolio life cycle are discussed. There are clearly inflection points where a shift in perspective from the enterprise portfolio to the more familiar project delivery is needed. These relate to the point at which the life cycle decomposes and attests to the recursive nature of any enterprise life cycle.

8.2.1.1 Multistreamed Life Cycles

The longevity of textbook models may have something to do with the complexity inherent in describing multiparty interactions such as those found in the enterprise pattern. The collective silence that we find on the subject is not simply a result of growing complexity. There is an attraction to a simple model. Presenting multiple interconnected life cycles can result in visual clutter. An argument could be made that once a life cycle is understood in its simple, singular form, then it is simply a case of aggregating multiple life cycles together in the execution plan. In that way, the model could remain simple and be a useful heuristic, whereas its realization as a project plan will necessarily be more complicated.

Armed with these rationalizations, managers within the project paradigm have persisted with traditional life cycle models, even when their avoidance of project dependencies regularly creates challenges that are out of scope for any individual project. Even system integration models like DoD-STD-2167A discussed earlier have been neglected because many consider the model too complex to apply in reality. Yet, clinging to simple models is only acceptable if we do not miss anything fundamental—so long as the reality is not significantly different from the model.

Complexity should not be allowed to swamp clarity– unless it is complexity itself we are trying to understand. Unlike traditional models, the enterprise pattern focuses on the issue of cardinality—the number of parties involved in the solution delivery—and relatedness. The many-to-many relationship between project and component teams, or more generally customers and suppliers, will inevitability result in more complicated life cycle models, but they will also more accurately represent today's reality. When looking to manage complexity in the enterprise, success will not arise out of the denial of complexity—it must come from a new level of abstraction that offers better management tools.

Animating the enterprise pattern gives rise to multiple intertwined life cycles shown schematically in Figure 8.3. Here the participants in a particular release are identified as three projects P1, P2, and P3. Assuming the same components are impacted as in Figure 3.2, we can expect three concurrent, multistreamed life cycles—each an instance of a system integration model as shown in the highlighted P2 project life cycle. The key difference from past models is that there are now multiple customers and so multiple programs executing in the same time frame. The relationships found in the enterprise pattern become the focus at two points: prior to the commencement of the individual component development in execution and before the integration phases at the end. These are the synchronization milestones that are also found in system integration models, the familiar fan-out/fan-in points that aim to ensure that there is a coherent design at the system level prior to allowing the different components to commence. Likewise, at the other end of the life cycle, components have to all be tested and delivered before they can be integrated to provide system level capabilities.

In system engineering terminology, these points represent key milestones in the program where baselines need to be established across all components prior to development proceeding. Focusing on the highlighted P2 project life cycle in Figure 8.3, we can get a close correspondence with DoD-STD-2167A; and the three broad stages reflecting the planning/design, execution, and final delivery of the P2 solution. As in the system model, there is a need for coordination between parties before moving from one state to the next. Without this and an agreed baseline being established, potential costly changes cannot be avoided. As a result, the component development that is undertaken at the execution stage is bracketed by the project (or system, to maintain a correspondence with system integration models) perspective and what are essentially design and integration activities. But unlike earlier life cycle models that can rely

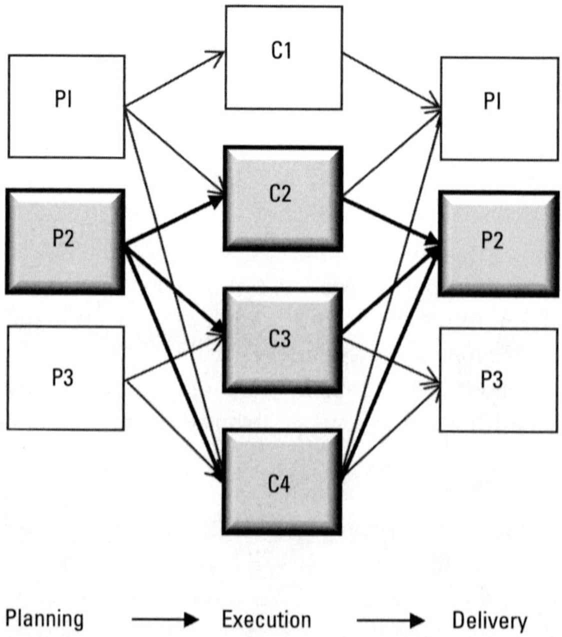

Figure 8.3 Multistreamed life cycles.

on a single point of coordination and control, a program manager or other customer proxy, Figure 8.3 clearly shows that some component teams can have up to three separate customers, raising the prospect of design and priority conflicts. These can be particularly debilitating since these are peer-to-peer discussions among projects that can be from different parts of the organization.

Hence, in the enterprise life cycle, as distinct from previous single-customer models, the baselines that need to be established are now "enterprise release baselines." Previously, each project team had to determine its own "design baseline" and OK it with impacted component teams. In the enterprise life cycle *all* design baselines for a given release should be agreed *together* to guarantee that there is no cross-project interference. This is clearly a new, enterprise level baseline distinct from the individual system baselines of the past. Focusing on baselines gives a different perspective to the need foreshadowed in Section 5.2, where planning and design was shown to be necessary at the release level and not simply the project level. While a project's solution architecture should always be reviewed in light of the wider enterprise architecture, the need for a release baseline makes clear that all participants need to agree on the release architecture previously described.

The need for these multiple layers of architectural governance may appear daunting, but these considerations arise naturally—either with proactive design intent, informal communication across project teams or, most problematically, as part of a surprise discovery late in the testing cycle. At least by recognizing the problem, we can attempt to better represent and share project information, improving the peripheral vision of the teams by establishing a common perspective of the release.

8.2.1.2 Family of Life Cycle Matrices

As we now consider the pattern in the temporal dimension, the resulting multistreamed life cycles can at first appear daunting. The release matrix, as introduced in the previous chapter, was shown to capture the key relationships in the enterprise pattern in tabular format with each cell of the matrix recording the allocated requirements of the impacted projects. Given this compact representation the matrix notation offers, we look to extend it in the temporal dimension.

The enterprise pattern has a specific incarnation for each release, capturing the projects and component teams along with their dependency relationships. As the change portfolios to be implemented in a particular release are selected, a unique release matrix is established, which persists throughout the life cycle—unless of course the scope of the release itself is changed. Hence, a family of matrices can be used to record the multiparty interactions at each stage of the enterprise release life cycle. And the previously discussed benefits of the matrix notation remain valid for each of these stages or disciplines that utilize this format.

As shown in Figure 8.4, pertinent life cycle information for each contract can be recorded in this collection of release matrices. The information provides a summary of what is traditionally unstructured and somewhat verbose documentation. The figure shows how the release progresses from the original allocated requirements through to the estimates, planned milestone dates, design information, test cases, and finally the version of the components. This last phase of the release reinforces the link with configuration management that was a motivation for a release-centric viewpoint from the outset. Indeed, where a release is well-planned and executed, we can expect a single delivery baseline for each of the components. That is to say, in a specific release, each column or component (see for example v_{14}, v_{24}, and v_{34} in the figure) should be met in a single design and component version. Such a scenario is not always possible and certainly not advocated, as it increases overheads associated with maintaining multiple versions of the C4 component. While common in software development, an enterprise example is the concurrent operation of different business processes for the same customer transaction—clearly a costly and undesirable scenario.

All phases of an enterprise release can be captured by such a family of matrices—clones of the original template that defined the scope of the release. The

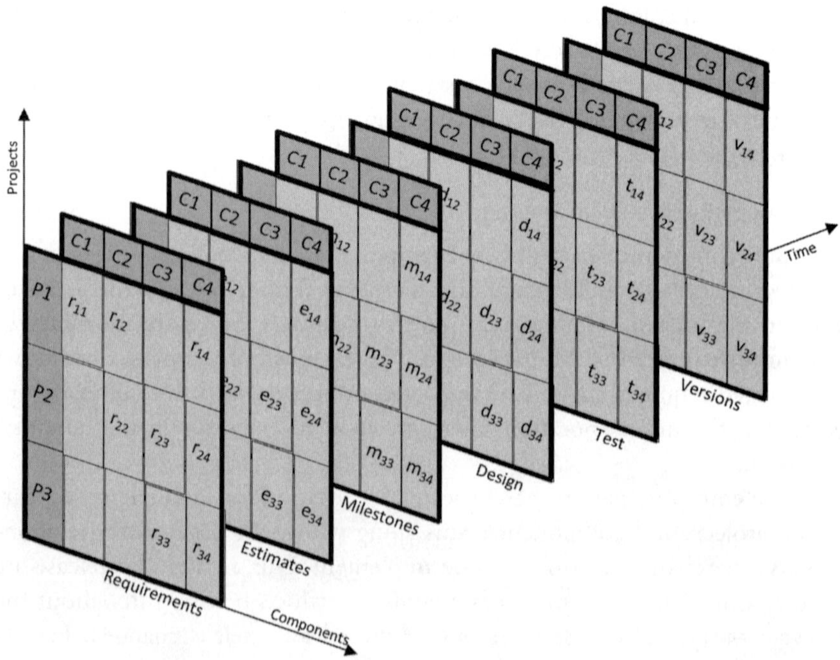

Figure 8.4 Release life cycle matrices.

key fact of who impacts whom can be discerned at a glance, as can the dependencies that projects have on component teams. There are cell contents, if there are impacts of that project on the specific component of the architecture. Once an impact is identified, that cell in all matrices can be used to describe how the original requirement is to be addressed throughout the life cycle. A longitudinal view of each requirement or contract is possible, acting as a form of traceability across the lifecycle and enabling different views of the portfolio: choosing the lens to view either a project (row), a component (column) or requirement (cell).

It needs to be stressed however that there is something more subtle happening than multiple instances of separate, waterfall life cycles being undertaken concurrently. The value of the notation arises where the implicit planning dependencies *between* requirements are able to be discerned by taking a holistic view of the release. These are not explicit in the matrix itself, but become evident when these matrices are queried to determine the knock-on consequences of any defect, change, or delay to a particular requirement in the release.

8.2.1.3 Synchronized Inflection Points

The enterprise pattern does not impose any time ordering on the relationships between projects and component teams. In an uncontrolled environment, which many enterprises would reflect, projects could randomly impact components of the enterprise's current operational architecture. This was the situation that we argued in Section 5.2 was inappropriate and risky in the enterprise.

Therefore, all timings associated with the selection and implementation of the change portfolio are imposed by release governance—coordination of these activities is management expediency. In establishing an enterprise roadmap, we made decisions as to the schedule, frequency, and rigidity of the release calendar. Within this framework, the scope for each release has to be determined—the projects and change initiatives that need to deliver capabilities in the release's time frame. Now we are at the point of discussing how the scope assigned to a specific release is best executed, and again best practices impose a temporal structure on the release life cycle in order to ensure a low-risk, quality release.

As in the original waterfall life cycle, sequencing of tasks in a release aims to achieve a consolidated deliverable prior to the completion of a phase. This helps minimize rework and change in subsequent phases but it need not be a rigid constraint. A risk assessment can make for a more flexible enterprise release life cycle. This argument goes back to the discussion in Chapter 2, when the direct (waterfall) path to a solution was shown to be more efficient (if not effective) than an iterative process, yet agility has its own merits. The more regimented we make the life cycle, the more efficient the construction phase—although this is at the expense of the analysis and design effort that must attempt to specify the complete solution before it can be built. Sequencing in the life cycle is therefore a trade-off that needs to balance the risk in one phase over another.

The desire to have a complete and agreed design before assigning work to component teams drives the first synchronization point shown in Figure 8.3. This is based on the view that it is more cost-effective to spend time agreeing and baselining the design, be it the solution or release architecture, than to have component teams commence earlier but run the risk of making adjustments as a result of design changes. As in the original application of the waterfall model, the latter concern is particularly relevant when the component teams are external agencies and changes are the subject of contract variations. Hence synchronization comes as a result of holding off component development until design has stabilized, from front-end loading, the experience that tells us it is better to get organized before having the teams work in chaos.

Likewise, at the end of the cycle the different components must be integrated together to provide the solution capabilities. Here there is less need for best practices to guide us as things simply will not work until all the parts have been put together. But while the need for an integrated whole demands the completion of all component deliveries—the last component to be delivered determines the release date—synchronization of those deliveries arises from the fact that it is not efficient to have teams deliver capabilities too soon. Early delivery of a component provides little benefit for the release. As in the print greeting scenario described in Section 3.1.2, having a team provide a capability before it can be usefully deployed represents suboptimal usage of those resources that could be better assigned to another activity. Ultimately, a component cannot be fully tested and accepted without confirmation that it will work with the other solution components, making early delivery irrelevant. As in manufacturing's just-in-time philosophy, it is more efficient to have all the component deliveries synchronized so that release managers can focus on the integration and test of the release.

Figure 8.5 applies a timeline to the three stages of an enterprise release life cycle to highlight the synchronization necessary to baseline the planning and design activities. The coordination needed across many parties involved naturally creates timeboxes *within* the release life cycle. The lock-step approach of the waterfall is difficult to avoid when dealing with this multiplicity, although individual teams have the discretion to develop their own plans within these timeboxes and optimize their resource utilization. So while each project undertakes its inception and planning phases independently, once it is selected to be part of a release, the solution architecture has to be completed, the component suppliers engaged, and their individual execution plans negotiated. As indicated in Section 5.2.1, regular timeboxed releases would see these negotiations confined to scope and costs given delivery schedule is fixed. Although, when major programs are included in the release, there may be a need to be flexible on the release schedule. Irrespective of the timeboxing strategy, these project negotiations must then be reviewed at the release level and confirmed to be compatible with the release architecture. The synchronization at the end of the release

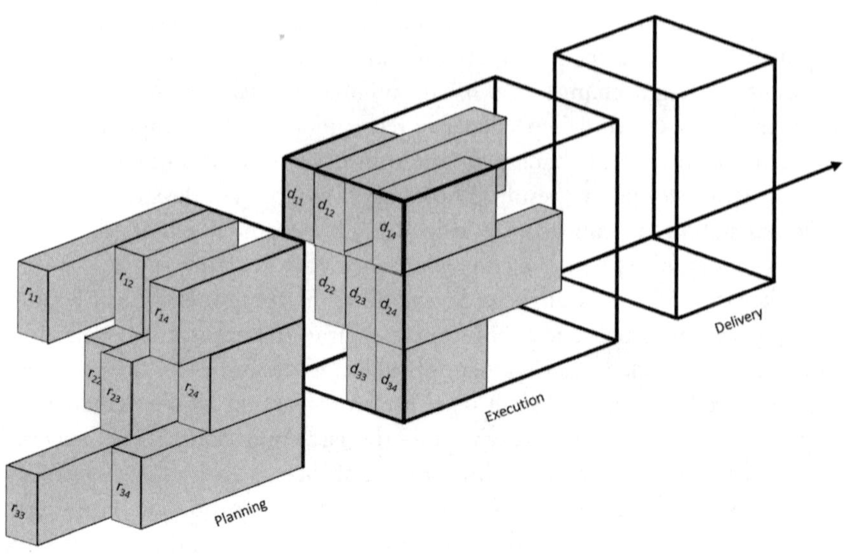

Figure 8.5 Synchronized execution and delivery.

planning stage captures the need to ensure design compatibility and lock-in the different customer–supplier contracts needed to execute the release. The projects must metaphorically line up at this release-level gate so when agreement is reached and the go-ahead is given, the entire portfolio moves to the next stage of component development.

Depending on local policies, the boundaries between planning and execution can be porous. Flexibility can be introduced with risk-based assessments of project activities, allowing some to start or finish earlier than others in the tranche. What the figure does not show is how a particular component supplier executes the different requirements. For example, the C4 component team has as their scope d_{14}, d_{24}, and d_{34} and they can plan the execution of these requirements to suit their own needs—so long as they complete within the release's execution timebox. Indeed, so long as component teams work within their timebox, they may have total flexibility in terms of the processes and tools they use for development of the component. This makes the release life cycle clearly a more general delivery life cycle where the type, medium, or processes that define the component are not relevant, so long as it is an agreed part of the release architecture and is delivered within its timebox.

This makes clear the distinction between the enterprise view and the component view, which is similar to the distinct system versus component perspective found in system integration models. The baselines at the synchronization gates represent transitions or inflection points in the life cycle. A shift in responsibilities takes place at these points. So long as project or system requirements are met, the component teams can act as a black box, both in terms of the process they use, as well as the internal design of their component. The perspective shifts with the hand-off from one team to another; for example, from the project team to the component development teams, or from a system integrator to the subcontractors responsible for the delivery of the components.

These inflection points in a multistream life cycle represent a critical stage requiring the greatest coordination and communication between stakeholders to ensure the transition from project to component perspective is efficient and risk-free. The fact that an enterprise release adds another layer of complexity can be mitigated by the increased abstraction of the release paradigm, supported by the compact matrix representation. Together they offer a framework for viewing a complex reality with life cycle matrices providing a simple tool to tackle the challenges at critical inflection points in the enterprise release life cycle.

8.2.1.4 Visualizing the Release Life Cycle

And while change and transformation represent a daunting challenge, enterprise leaders demand simple solutions. When we encounter such complex challenges in system design, we look to rearchitect or reengineer the system to make it manageable and at least appear simpler. Likewise, we need to shield enterprise leaders and managers from complexity and ERM attempts to do that by taking the release as the abstraction rather than simply the projects as the abstraction.

But in describing the multifaceted nature of the enterprise release, words alone fail to adequately capture what is actually taking place. This is especially true when the words we use are overloaded with meanings that have been ascribed to them by the traditional project paradigm. So when a textual description becomes difficult, we need to resort to the universal language of images.

Figure 8.6 offers a visual representation of the enterprise release life cycle. The figure shows a single release with its end to end life-cycle phases. The project activities are shown as horizontal blocks (corresponding to the rows of the matrix) impacting solution components, shown as vertical columns (as in the matrix), while synchronization points requiring release governance are shown as full-sized panels. The figure adds indicative development phases that fall within the planning, execution, and delivery timeboxes. These are only examples dependent upon local conventions, however—the release matrices detailing each phase can be considered as cross-sectional slices taken at specific points in the life cycle. The individual cells in these matrices are not shown but would capture

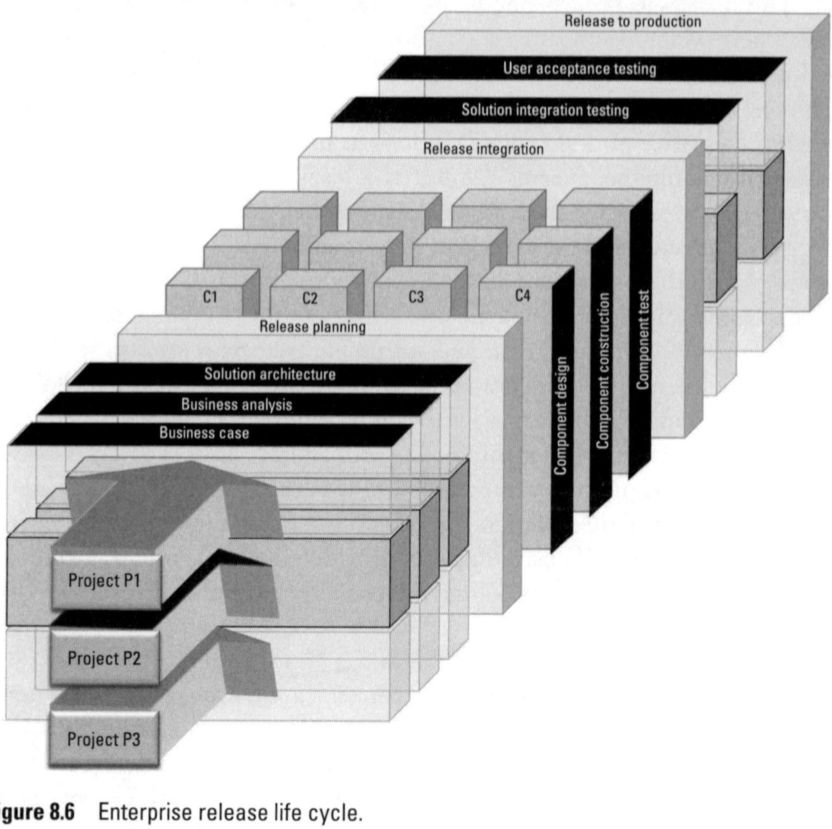

Figure 8.6 Enterprise release life cycle.

the points at which projects impact on components, their contents describing (initially at least) the project requirements allocated to that component.

Taking an end-to-end view of this visualization, one might imagine a "fly-through" of the release life cycle from start to finish and in this flight of imagination, have to negotiate the inflection points by turning or twisting to switch between the project and component perspectives. Starting at a project, a twist is required at the first inflection point, taking us from the project to component perspective while at the second we twist back from component to project perspective, or more specifically to the solution that a project delivers. The re-orientation required at each inflection point corresponds to the need for release level governance—first at the release planning/architecture stage and second, the integration/test stage prior to delivery. If the fly-through is to be accurate, then it should include a simulated delay at the inflection points to reinforce the synchronization, or at least the alignment needed in order to ensure all parties make a coordinated transition—ensuring solution architectures are compatible, requirements are clear, and commitments are agreed before commencing the next stage. In other words, each party has to ensure that their role in both project and component dimensions of the matrix are locked in.

For all the abstractions above, the release life cycle is unavoidably more complex than previous life cycle models—and complexity is an anathema to management. The inability to convey the problem simply may contribute to their poor acceptance in practice. But with the lack of some coordinating activities, project success will remain elusive in the enterprise.

It is to be hoped that ERM, with its more appropriate enterprise-level abstraction, provides a framework that can provide management visibility to the enterprise leadership team, even if complex beneath. With the aid of the release matrix, with its support for multiple viewpoints and time-lapse visualizations, the interactions between multiple parties in the release life cycle can at least be better communicated and understood.

Recognizing the limitations of words, additional visual aids can be accessed online at enterpriserelease.com. Available resources include presentation slides and animations demonstrating the fly-through aspects of the enterprise life cycle—readers are invited to join the ERM discussion.

8.2.1.5 Matrices as Scalable Documentation

Much as the project team needs to work cohesively, there is now also a need to coordinate the work and work products of the extended team targeting an enterprise release. The matrix notation, in compactly capturing life cycle information, has the potential to improve communication and information dissemination from a release-centric perspective.

Projects traditionally develop a suite of documentation from requirements, solution design through to test and implementation. This remains true in most organizations today, and the volumes of documentation produced can often bury project personnel and business sponsors in the detail. In attempting to avoid document overload, agile projects try to adopt a less formal approach with the simplistic belief that ultimately "the software is the documentation."

The intriguing possibility that is being researched and prototyped (again, see enterpriserelease.com for more examples and templates) is to utilize life cycle matrices to record, or at least reference, information across the release. For smaller, agile releasing the cells of the matrices could potentially be implemented simply by a spreadsheet, sufficiently capture the different pieces of life cycle information. In larger programs, the cells in the matrices could link or reference artifacts that follow the organization's documentation approach.

As an example, the requirements denoted in Figure 7.9 as r_{11}, r_{12}, and r_{14}, represent the allocated requirements for project P1's solution architecture. In the traditional project methods employed by most organizations, these would take the form of one to three requirement specification documents that had to be provided to the component teams. Similarly, d_{11}, d_{12}, and d_{14} would represent project P1's more detailed solution design deliverables for components C1, C2, and C4, respectively—now more likely to be three documents. The number of documents can proliferate with interface documents, development plans, test cases, and so forth. And that is simply for a single project in the release. When the documentation from other projects is included, navigating this "library" to find the latest, pertinent information can become problematic for the release team.

It is ultimately those responsible for the final delivery phases that have to accommodate an avalanche of not-always-clear project documentation. At this phase, the need for release management is self-evident. Those who have had to process this mountain of information in order to develop test and implementation plans would appreciate the need for a more succinct and compact form of "documentation." And when it comes to agile projects in the enterprise, the opposite applies—there is often a dearth of sufficient information for enterprise testers since agility presumes the project is self-contained, stand-alone textbook.

At present, there are few alternatives for capturing interteam communications. Projects either resort to the traditional documentation deliverables that formally specify requirements and designs, or where agile methods are preferred, rely on more ephemeral story cards written on Post-It notes, informal communications, and standup meetings, such as the Scrum of Scrum, conducted *between* collaborating agile teams. However, one of the problems encountered in many larger organizations is that there is a mix of approaches adopted by different teams. The challenge in such instances is to communicate between these teams using a shared lingua franca, or working language, to bridge the needs and preconceptions of the different approaches used.

The family of release matrices offers a succinct and self-organizing documentation format. It could simply be a filing system for the release information, allowing participants to navigate potentially separate project or component team repositories, or it could actually form the basis of a simple tool. The latter holds the promise that a lot of the coordination and management complexity associated with the release life cycle could be alleviated. As the delivery milestone approaches, ensuring that the correct version of the requirement is being tested can significantly reduce confusion and time-wasting—a pointer to the latest version of the relevant document can be extremely valuable.

In trials of the matrix notation, the separate specifications used in requirements were deconstructed and reformatted into spreadsheets corresponding to release matrices. The results showed that the original requirements and design documentation could be more conveniently represented in this structured and concise notation. This suggests that the matrix notation has possible extensions beyond being simply a model and has the potential to be a means of structuring all documentation and communications relating to a release.

8.3 Strategic Enterprise Planning

Strategy is not likely to be executed in a single release, no matter how transformational or significant the planned changes may be to the business. The discussion to date has described a singular enterprise release, but strategy is more likely to be executed over a series of planned releases that we have termed the enterprise roadmap. When planning and designing an upcoming release, those initiatives that do not get approved can be considered as candidates for future releases.

Figure 8.7 shows the release life cycle, in the context of the enterprise roadmap showing, literally, the shades of gray as the strategic plan stretches into the future. This illustration shows three planned releases with requirements allocated, although perhaps not baselined or agreed. Even future releases can be seen as a work in progress and may have skeleton staff assigned to perform scoping and feasibility investigations. These activities prepare for the execution phase and finalize the release scope in readiness for the component teams who are only able to commence work once they conclude the previous release.

Just as life cycle models have evolved to recognize the constancy of change and adopted more agile and iterative development processes, so the enterprise release life cycle can become more effective in delivering strategic outcomes, when the planning and execution processes are connected seamlessly. ERM views the release as the natural nexus between these two sometimes separate worlds. It is the manner in which we arrange and prepare for the releases that can provide new efficiencies.

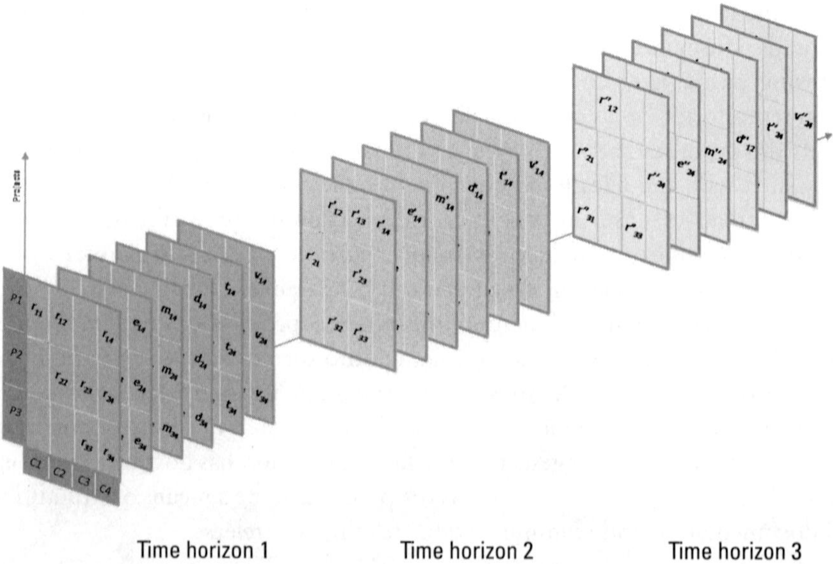

Figure 8.7 Enterprise roadmap time horizons.

While the enterprise and their architectures become complex and the organization unyielding to change, what is flexible and can be agile is the planning of strategic change. The enterprise architecture unlike software cannot be easily branched, versioned, merged, and refactored[1]. All those things ARE possible and easier to achieve in planning—it is only when plans are implemented that the inflexible arises. The real potential that a release-centric approach offers an enterprise is to become efficient in its planning and to use agility to reduce management overheads and streamline the logistics of enterprise change.

8.3.1 The Enterprise Roadmap

Guiding enterprise evolution requires making decisions that go beyond the timescales of a single release. While strategy may call for plans going out 5 or 10 years and beyond, in today's competitive business, planning may only be relevant over a period of months rather than years. Irrespective of what constitutes the strategic "pulse" of an organization, the release as a planning construct remains pertinent. In fast-moving industries especially, the identification and continuous elaboration of an enterprise roadmap readies the enterprise for the execution phase, laying the foundation for successful delivery.

As was stated at the outset in Chapter 1, a release can come into existence long before IT or any delivery group gets to know of it. They may not be called

1. Components can of course branched, resulting in duplicated systems and processes that are costly to operate and eventually need to be consolidated.

releases at this point in time but the release-centric paradigm clearly can link the planning and execution of strategy, especially in organizations which have a continuous stream of changes.

Relating strategy to releases gives an ephemeral vision substance with a clear set of delivery dates that can be communicated across the organization as necessary. Strategy becomes a real and achievable action plan for the enterprise that can then begin to adjust behaviors and prepare for the changes.

But to take ERM and the release paradigm to that level requires more than simply placing release dates on a calendar, although that in itself can be quite powerful in setting targets and expectations. The new dimensions that the enterprise roadmap introduces to the psychology of scope negotiations are trust and time. For customers and suppliers to commit to such a roadmap, they must believe it to be a reliable indication of what lies ahead—in a release-centric approach, we are asking stakeholders to buy futures and to trust in the promise of a future state.

8.3.1.1 Planning the Enterprise Backlog

The enterprise roadmap lies out in a time line, the points at which a fully supported enterprise release can be expected to be delivered—the timeboxes to be filled with prioritized changes. This takes us full circle and back to the Enterprise 2.0 analogy that viewed the enterprise as a series of planned releases. While there may be other smaller releases across a large enterprise, the roadmap highlights the strategic release that aims to transform the organization over time.

And time is an important consideration in release planning. Prioritizing projects or specific scope items that range across the enterprise presents a significant challenge with the competing and often conflicting demands being impossible to reconcile and balance within a single release. While most customers want to see their requirement implemented as soon as possible, the future presents an opportunity for negotiating compromise. Scrum terms those requirements that cannot be implemented in the current sprint (in our language, a release) a backlog, and project scope follows the same process. The enterprise roadmap therefore provides temporal placeholders for the enterprise backlog, comprising the initiatives (or parts thereof) that were not able to gain approval but may have relevance in the future. Establishing such a backlog at the enterprise level provides some assurance to sponsors disappointed that their initiatives were not approved in full, can be reliably expected to be executed at an agreed point in the future. While clearly a compromise position, the value and trust a customer has of such "promises" to deliver their requirements in the future, is directly related to the confidence they have in the published enterprise roadmap.

There are no fixed rules to establishing such a roadmap, and business conditions should determine what is most appropriate. For some organizations, determining the frequency and timings of the releases in the roadmap can be a somewhat arbitrary decision. The decision may be based on the organization's capacity to absorb change, the pace of change in the industry, time-to-market considerations, or the funds available to fuel change. For other organizations, there are external market factors that drive the release calendar, including financial reporting cycles or the tax year. However, there are clearly implications that have to be considered including release logistics, staffing, and release efficiency. Too-infrequent releases are likely to be less efficient because release teams suffer resource turnover, lose critical knowledge, and then have to remobilize. Increasing release frequency improves these considerations with the consequence that operations and business users suffer greater disruption. On the other hand, for those with a vested interested in the changes, as opposed to operational stability, more frequent releases mean less of a wait when initiatives get pushed out to future releases.

While infrequent and ad hoc releases may suit some organizations, it is the repetition associated with frequent releases that improves the efficiency of release. Further, when a regular release cycle, the so-called release-train metaphor is adopted, it opens the possibility of sequencing release activities more intelligently. Rather than considering releases in sequence and only starting to think about the next release at the completion of the preceding one, advance preparation for future releases offers the means to streamline delivery. Specialist roles can move from release to release, working across the different time horizons, to perform pre- and postcomponent development activities. Efficiencies can arise from having these teams prepare the groundwork for future release or, at the other end of the spectrum, integration and test teams prepare for the delivery. Crucial release governance activities can therefore be conducted well in advance of the release's execution, mitigating the schedule impact of added processes that focus on the release's design integrity and architecture.

The efficacy of working concurrently on multiple releases in a production-line-like manner suggests that a regular, timeboxed release train is the most appropriate and efficient means of delivering a change. It suits enterprises experiencing, or desiring, rapid and continuous evolution. The routine of regular releases at known intervals essentially gears the enterprise for changes. The efficiency benefits arise as a result of the preparation and groundwork conducted before and during a release in order to lubricate its passage through the life cycle. To do that, a release-centric approach invests in the processes and shared resources needed to facilitate the release life cycle rather than growing the number of project roles.

8.3.1.2 The Role of Release Architect

The difference between a project and a release perspective can be described with an analogy. If changes had to pass through a funnel to be implemented, then

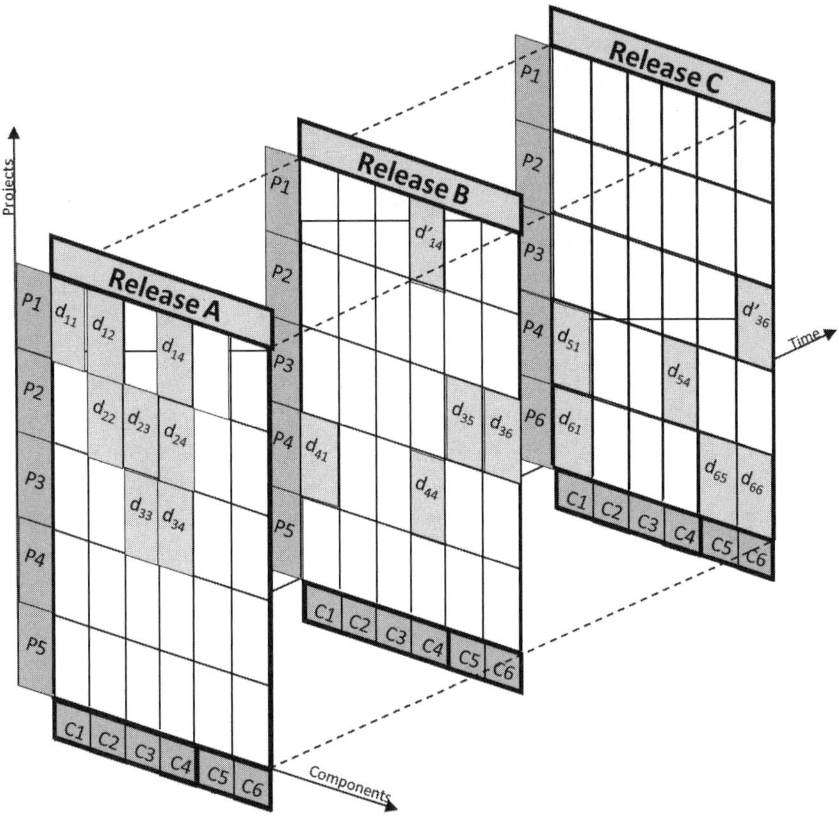

Figure 8.8 Enterprise roadmap showing different parties involved.

the project view is to push down on the funnel to get more changes through, while the release-centric approach attempts to make the funnel's spout larger to increase change throughput. A comprehensive ERM framework can ensure reliable delivery of strategy that avoids excessive management overheads and reduces the risk of "breaking" current operational systems. To establish such a framework requires we must step outside of a single release and consider instead the enterprise roadmap.

It is in this longitudinal view that considers the sequence of planned future releases, essentially potential to-be versions of the enterprise, that we have the opportunity to prepare and plan outside of the pressure and chaos of the current release. Here is the realm of the release architect, a new role that addresses the enterprise backlog and prepares the path for the release. If this appears to be simply restating the role of an enterprise architect, it is not. The points of differentiation include:

1. The role acts as the custodian of a specific future release, defining its boundaries in line with corporate budgets and resourcing;
2. A broad release theme is developed that aims to satisfy an area of strategic concern to provide a unifying design purpose;
3. The backlogged strategic initiatives and requirements that are assigned to the release drives the planning so it is business aligned;
4. The role stays with the release as its scope is elaborated and takes on the mantel of design authority for the release as the execution stage approaches.

With these responsibilities, the release architects can be a point of focus for a release, working with the leadership to flesh out their vision and adjusting it as execution approaches. As the design authority for the release, they must facilitate the design and (re)estimation of initiatives and requirements assigned to the release and ensure that the proposed solution architectures are compatible with the emerging release architecture. Such foresight and planning may appear counter to the fast-paced reality of today; however, a strategic focus need only account for part of the release's budget aimed at longer-term strategic capabilities. The ability to support more tactical or newly identified initiatives exists so long as the resources capacity for that release is not exhausted. And when that point is reached, the execution portfolio can be reprioritized with those initiatives not able to be placed, where relevant, on the backlog.

The strategic architect, who may be supported by a team in larger organizations, has accountability for their assigned release. Working towards a distant time horizon, the release landscape and its participants may be quite different from that of the current release. And the matrix notation needs to be extended to offer the capability to include multiple releases that can accommodate future projects and the components of the enterprise architecture that they impact. Clearly, many of the parties identified are not participants in every release, and when considering a single release in isolation, the notation can be collapsed to show just those participants directly involved.

The enterprise roadmap shown in Figure 8.8 represents the planned delivery of the strategic vision as three enterprise releases. It shows the axes extended beyond Release A that we originally considered, to accommodate those project-component impacts that are anticipated in the releases identified. Each release has a unique scope and as that is delivered and warranty obligations met, the release architect can be rotated onto a future release that has yet to be planned. As a new release now comes into currency, it must validate the current baseline that it has to be built upon. This would largely be defined by the preceding release—except for those change programs that did not participate in the enterprise release. In the rotation of responsibilities, each release architect has provided the focus of design decisions from planning through to execution, so helping to avoid the wasteful reinvention that can arise between business case and execution.

The critical role of the release architect provides for periods of intense activity followed by longer periods of preparation. It emerges naturally from a release-centric perspective and the alternation of planning and execution helps ensure that a balanced perspective is maintained. While these new roles may appear to be an overhead, they are first an investment in the execution of strategy that by definition is warranted, and second, they can improve the capacity of a release while reducing its delivery risk.

The appointment of a design authority and custodian of a future instance of the enterprise helps lay the groundwork for regular enterprise releases. While the task of getting oversight of the entire enterprise can appear daunting, in reality tools based on the matrix notation can assist with the coordination of planning and design activities, making the release architect essentially the keeper of the matrix. And the role ultimately provides support to the leadership team to ensure that their strategic vision guides the evolution of the enterprise—and identify where it does not.

References:

[1] W. W. Royce, Managing the development of large software systems, *Proceedings of WESCON,* 1970, pp. 1–9.

[2] *A Spiral Model of Software Development and Enhancement.* Boehm, Barry. 4,

[3] The V Modell XT, Development Standard for IT Systems of the Federal Republic of Germany, The Commissioner of the Federal Government for Information Technology (originally 1992).

[4] DoD-STD-2167A, Defense Systems Software Development, 1988.

[5] ISO/IEC 12207, Systems and Software Engineering—Software Life Cycle Processes, 2008.

9

Practicing Enterprise Release Management

Even if there is only one possible unified theory, it is just a set of rules and equations. What is it that breathes fire into the equations and makes a universe for them to describe?
—Stephen Hawking

Transformation and change in the enterprise is arguably the most difficult challenge of our time. The sophisticated systems and architectural complexity behind simple, everyday business transactions such as Internet banking or credit card payments are easily overlooked in our blithe adoption of these conveniences. Even as enterprises attempt to adjust their business models to harness and adopt the latest technological innovations, they have to be built on existing architectures that are simply too critical to replace without risk to business operations. To embrace the latest Web 2.0 social media features and link to customers via Twitter and Facebook often means having to interface into legacy back-end systems. Each successive wave of innovation adds another layer of complexity which, because of its incremental nature, is too easily ignored.

Only occasionally, as in the Y2K crisis just over a decade ago, businesses are reminded that that so much of what is taken for granted is in fact made possible by sophisticated systems and processes that are not well understood or managed. In the case of the so-called millennium bug, it was society's reliance on information technologies that became clear. The difficulty of remediating a relatively simply date format embedded in the most malleable medium of software highlighted the change management challenges with business' increasing reliance on complex and interdependent IT systems.

Today, the inevitability of change and increasingly globalized competition has introduced new demands on management. Businesses are caught between the need to be flexible, dynamic, and agile even as they maintain their investment in people, processes, and technology. A new challenge to leadership arises when business alignment and transformation clashes with the need for continuity, stability, and controlled change necessary to maintain complex enterprise architecture. Where these fundamental forces are not balanced, the result can eventually lead to dysfunction as a fractured architecture grinds business down. The best hope we have of managing the debilitating change and complexity cycle is to better govern change in the enterprise, recognize the need for a measured management response, and establish processes that are responsive to complexity.

What motivates ERM, or in terms of Hawking's quote, what breathes fire into the proposed matrix model, is the need for new management paradigms. There is a gulf between textbook approaches and the reality managers and practitioners face in reality. The modern devolved and distributed enterprise is nothing like what best practices would lead us to expect and even as industry bodies are working to address the emerging need for portfolio governance, these initiatives often lack a unifying paradigm. The release-centric perspective goes back to the future to present a new management synthesis that hopes to stimulate new, innovative approaches to delivering enterprise change and transformation.

While the conditions present in today's businesses introduce new management challenges, they are eerily familiar to the problems encountered in complex systems and software engineering. The enterprise release offers a high-level abstraction of the release management discipline in product development. Thus it is underpinned by both research and practical experience, albeit in a different domain. Suitably scaled, these fundamental principles result in ERM, which therefore offers a unifying paradigm suited to the increasingly distributed business models and complex supply chains found today.

In this final chapter we review the fundamental characteristics of ERM and look at ways in which it can be applied in the enterprise. The implementation steps needed to adopt the ERM in the enterprise are identified and practical approaches for applying the key concepts are described. In addressing a recurring pattern, these techniques can offer benefit at different levels in the organization. The opportunity to bring together the myriad disciplines around the release paradigm offers greater role clarity and streamlines delivery. And while achieving a more agile approach to delivering strategic change requires enterprise-level adoption of the key principles of ERM, the first steps in that journey are simple and practical, requiring little effort other than a reframing of existing approaches. If the release-centric paradigm is appropriate in the organization, then it should not be difficult to implement. At a fundamental level, if something is too difficult it is wrong.

9.1 ERM as a Unifying Approach

After the long traverse it pays to overview what ERM entails and consider how the different concepts introduced can be applied in the enterprise. At the heart of ERM is the enterprise pattern, which is a direct consequence and manifestation of the twin forces of change and complexity. In plotting these forces on two separate axes, the fourth quadrant, where both of these forces were in evidence, corresponded to a weak spot for management. Matching the finding of the Bain study, this quadrant was shown to present opportunities when business alignment and well-oiled management could both be achieved.

However, these conditions are not unique to the enterprise and the ERM is broadly applicable across a number of domains. It is not just the enterprise that faces the challenge of managing the enterprise pattern, which also described the management challenges found in product line architectures [1–3]. There, as in the enterprise, new product features impact a common platform architecture that must evolve in the face of multiple demands from the product groups.

There too the multiplicities of customer and suppliers relationships are bound together by the economic value, in this case of sharing the platform architecture. In product lines, dynamic tensions arise from the strategic need to keep that architecture relevant and reusable by all parties, even as they each are motivated by their separate customer's requirements.

As the enterprise pattern is an example of a system or systems, ERM can have application as a generic delivery approach. The wide applicability of the pattern is indicative of the sophisticated, multidisciplinary solutions that are delivered today and transcend any simple categorization. Its generality requires new release-centric terminology that has yet to be fully developed, with the current management vernacular steeped as it is in the project paradigm. What the examples illustrate is that where agility and responsiveness is demanded of complex architectures, ERM and the underlying release paradigm offers a synthesis of existing disciplines and a unifying management approach to solution delivery.

9.1.1 The Applicability of ERM

Louis Gerstner, Jr., who headed IBM through the tumultuous years after the introduction of the personal computer, used the metaphor of an elephant dancing to represent the agility and flexibility needed of the enterprise in a competitive market [4]. In essence, ERM provides a new perspective on how to successfully execute and deliver a business strategy—or how to make an elephant dance. While the tune, or strategy, may change arguably the ability of the organization to deliver and adapt to change, is the key to sustained competitive advantage. And the first step is to understand the elephant in question.

The matrix notation and it extension along the delivery life cycle provides an aggregated, holistic perspective of an enterprise's evolution. But capturing both the customer and supplier perspectives, the notation can better connect planning and execution and negotiate the inflection point between the two perspectives. It offers a management dashboard for executives seeking to understand the status of strategic execution and its reliance on parties both internal and external to the organization. An important distinction to make is that ERM takes a strategic view of the entire enterprise change portfolio rather than the individual programs or projects that deliver aspects of its strategy. By taking a unified approach to enterprise change (suitably filtered to offer a strategic perspective) and monitoring those component teams who are charged with its execution, delivery risk can be lowered while managers have improved visibility of delivery status.

This picture of a top-down, executive view of enterprise transformation and change may appear reminiscent of a sales pitch for any portfolio management tool. But it is clearly distinguished from existing tools in a number of key ways to be discussed below. And to further remove any suggestion that ERM is yet another silver bullet proposed to solve all our problems in solution delivery, we will identify a number of "relevance tests" to ensure the symptoms that an enterprise suffers are amenable to the ERM cure.

9.1.1.1 Validating the Pattern

Even where there is evidence of the enterprise pattern in different organizations across different industries, the key question that has to be asked of any organization seeking to implement ERM is: Do we have that problem? Is the pattern evident in the enterprise at this point in time, or is it likely to emerge within the timescales that the strategy is being developed?

Clearly from the change–complexity quadrants presented in Chapter 2 and the Bain study discussed in Chapter 7, there are organizations that appear to not suffer the enterprise pattern. However, the Bain study showed that the small percentage of organizations that occupy the fourth quadrant can derive significant business benefit from balancing alignment and efficacy. The challenge of sustaining this balance suggests that organizations may not embrace this opportunity and instead withdraw to a singular focus—managing for either change or complexity. There is some evidence for this view in the regularly restructuring that organizations undergo, alternating between the allure of agility and alignment versus the more centralized and efficient shared services model.

Irrespective of the current structure adopted or even the expectations and desires of the leadership, there is an indisputable reality that can be discovered with the appropriate analysis. The following features are the distillation of the enterprise pattern and so can be used to gauge its relevance in the enterprise.

The degree to which the pattern is in evidence determines the applicability of ERM and the likelihood of it improving strategic execution and delivery.

Concurrent Multiparty Change

The number of change initiatives in the enterprise provides an immediate indication of ERM's potential relevance. How these changes are managed or organized can vary but the mere existence of concurrent projects emanating from across the enterprise suggest the need for unifying the portfolio and applying ERM.

As each project attempts to manage the textbook, two-party relationship between customer and supplier there can be a proliferation of project and change managers. As requirements are allocated to different component suppliers the number of these relationships multiply, and with them the number of managers required. Competing for priority these intermediaries personify the problem of communications and so the first signs that ERM can help is the coordination across the different parties involved in delivering change.

Applying the project paradigm in a vain attempt to impose control over supplier relationships can see personnel assigned to key interfaces and relationships. And in response, the supply-side negotiations likewise can demand component suppliers establish a dedicated point of contact for larger programs. So each row and column of the release matrix, and sometimes even individual cells, need representatives to address the demand and supply challenge.

Many organizations have successfully concealed the reality of the enterprise pattern beneath management layers using either internal or external proxies. Internal groups can include IT departments and shared services groups, while externally there may be outsourced partners or vendors who essentially act as systems integrators. But while this can provide localized control for the teams, the wider need to manage and optimize the delivery of enterprise priorities goes unaddressed.

It is not individual team performance that we should seek to optimize but the more elusive enterprise performance. A deeper analysis of the organizational structures of project and component teams can be valuable in identifying the management tiers. Excessive management levels are costly and can make it difficult to determine the reality on the ground, often giving rise to inefficient and highly politicized organizations. In such circumstances ERM can at least expose the difficulties and ensure that project and component teams address the reality of demand and supply instead of having expensive resources lobby for preferential treatment that may be against the wider interests of the enterprise. The efficiency of ERM is that such governance decisions can be made outside of, and ideally before, the evident pressures of the execution phase.

Shared Dependency and Impacts

Simply having numerous projects in the enterprise that are reliant on different component teams does not in of itself constitute the enterprise pattern. If the multiplicity was able to be separated into completely independent delivery streams with no overlap or interaction, we would simply have a portfolio of disparate initiatives to manage. What gives the pattern its interest, and its complexity, is that these initiatives are interrelated due to their reliance on shared components and suppliers. It is the intertwining of initiatives and the resulting dependencies that are responsible for the nonlinear system behaviors exhibited in planning a change portfolio.

While a review of the solution architectures of each initiative in the portfolio would identify any shared architectural components, there can be a quicker means of identifying high-level dependencies. As recognized in Conway's law that considers architecture to be a reflection of communication and organizational interactions, it can be more instructive to review the relationships between participants in a release than conduct a detailed architectural analysis. Simply determining those teams that are involved across several initiatives can quickly provide an indication of the core capabilities that cause planning interdependencies. Shared services groups in the enterprise are a recognized impact point; however, there can be dependencies across the different dimension of the enterprise architecture, including:

- The reliance on critical resources that require initiatives to be planned around their availability. This familiar planning constraint now transcends individual resources and is often outside a project's control as it relies on the cooperation of managers who control access to specialist resource pools—whether internal or external to the organization.
- Architectural dependencies that arise when several solution designs place demands on a shared component. This is the most visible and clearly identifiable technical dependency that is familiar in IT but not always understood or made visible to the business. With the increase utilization of packaged solutions, outsourcing and offshoring these dependencies can impact vendors and partners that become integral to delivery.
- Impacts to business processes which, like the technology architecture, have to retain design coherence and be as simple and efficient to reduce costs and processing errors. Such processes are increasingly exposed to customers and can involve third parties that become part of the business value chain.
- Changes have to ultimately be acceptable to end users and even where initiatives meet all the other criteria, they can incur resistance. There may be difficulty absorbing the volume of changes or adjustments to

roles and responsibilities may demand careful negotiation and transition planning before they can be adopted.

ERM has a role to play where such dependencies can be identified and help make them explicit. These can manifest at different times in the release life cycle with impacts on enterprise resources serving to bind what might otherwise be independent initiatives into a release. An analysis of past "unexpected delays" can help identify common dependencies that can then be explicitly managed in the release matrix and mitigated in future projects.

Distributed Control

The enterprise pattern is most likely to be evident in flat organization structures because it relates directly to management's span of control. When under the control of a central authority, parties that have conflicting priorities can be readily brought in line. However, where the parties are more loosely coupled agents, each with their own interests at heart, there is a need for more collaboration and negotiation to achieve an acceptable compromise.

The pattern essentially identifies the need for federated or collective decision making across autonomous teams be they separate business units, departments, and even organizations. We find this where parties behave as peers with no clear escalation path, or one that is not able to adjudicate on the problem. This in fact is what makes ERM amenable to an agile management approach; it provides the greatest benefit to those organizations that do not desire, or cannot afford, the significant management overheads associated with control. Yet, the pattern can exist even where there is central control with the role of ERM and its associated tools shifting to provide a decision-support tool for management. In other words, while the need for ERM may be more evident in distributed organizations, however, the tools are method-agnostic and can be beneficial in both environments.

Increasing specialization and reliance on external partners in the supply chain results in a diminution of control. Each party has their own performance goals, and cannot be directed to change their architecture or business processes. Without direct control, management must now rely on tools such as the release matrix that are less intrusive but can clarify the points where decisions are needed and the implication these decisions may have on other parties.

Continual Change

Even where the pattern is evident in an organization, there has to be value in rolling out ERM in the enterprise. As in the recognition of the release as a life cycle process, it is only when change portfolios are regularly executed that the full worth of a repeatable process becomes evident. Not that the application of

ERM for a single release is not viable; larger programs commonly establish their own project office and document their own processes. But clearly the benefits of an established ERM framework comes into its own where there is a need to support the execution and delivery of a change portfolio on a continual basis.

That indeed was the premise behind the enterprise release paradigm: that it provided a reliable delivery mechanism irrespective of the scale of the initiatives that go into that release. The anticipation of significant changes resulting from market competition or a changing regulatory environment can be the catalyst for introducing ERM into the organization. However, any organization where the number of proposed initiatives exceeds execution capabilities will find value in ERM as a framework for managing its enterprise change portfolio and reducing delivery risk.

9.1.1.2 An Opportunity to Unify Management

Identification of any of the above symptoms in an organization suggests that ERM practices have relevance in addressing their challenges. Yet there are other advantages to the generalization that is inherent in the pattern. In particular, the matrix notation can provide a means of unifying the disparate technical and management disciplines that operate across the delivery life cycle. While each of these specialist perspectives are well intentioned and share the goal of working to the same strategic objectives, they lack a shared representation and can sometime interfere with each other, giving rise to process inefficiencies. This is the argument used by advocates of agile methods who seek to reduce the emphasis on specialist disciplines and instead rely on experienced practitioners who possess a range of skills.

As most current disciplines and best practices are largely based on experience they lack a shared theoretical framework; we have seen how the very fundamental definition of a project itself has been varying over recent times. These practices were never “designed” and instead emerged to fill real and perceived gaps found by practitioners largely in the days of the textbook and integration pattern. For this reason we cannot become too precious about these disciplines and they themselves can and should evolve over time. The risk of a highly successful paradigm like the project is that once it becomes the accepted orthodoxy any challenge to it can be viewed as heresy.

The good news is that adopting ERM and a release-centric approach at the strategic level does not require a complete overhaul of the current project paradigm; if it did it would most likely be doomed to fail. Instead, we recommend a more incremental approach adoption that meshes in with existing management structures and approaches while yet distinguishing and clarifying the different emphasis needed for ERM.

While the focus of this book is on managing change, both transformational and evolutionary, in the modern enterprise, the basic principles are borrowed heavily from software development, systems engineering, and product management. Even as ERM challenges much of the orthodoxy surrounding the prevailing project paradigm, it is a generalization that does not refute any of its fundamentals. Instead, ERM attempts to extend, translate, and apply its core principles so they are relevant in the modern business environment.

The ERM can therefore be seen as offering a unifying framework for the different management disciplines and "connect the dots" between the different industry standards and best practices. Familiar touchstones that the ERM and its matrix notation helps contextualize include:

- Project, program, and portfolio management representing the activity dimension with the change portfolio being the set of rows in the matrix notation;
- Enterprise and solution architectures representing the enterprise assets such as the technology and business components that make up the columns in the matrix;
- Change and configuration management, which addresses the temporal dependencies and possible combinatorics that must be agreed and baselined to enable the different component streams to deliver the required capabilities.

At its kernel, ERM links the project dimension to the architectural dimension integrating these previously disparate disciplines. As one manager describes it, ERM is a mashup of PM-BOK and ITIL. It applies traditional system engineering and configuration management principles to the enterprise, extending the traditional, one-off release management to be relevant across the end-to-end solution life cycle. Therefore, ERM represents the scaling of familiar disciplines to the strategic enterprise level. At a fine-grained level, the benefits of an integrated application life cycle is evident in software development where, with the benefit of SCM tools, changes are related to software artifacts and continuous integration enables regular, low-risk releases.

9.1.2 Time for a New Paradigm

The evolutionary picture of delivery patterns presented in Chapter 7 shows the enterprise pattern at the latest point in the trajectory that is defined by modern businesses incessant demand for change in the face of growing enterprise complexity. These conditions are unlike those of the past, when it was easier to focus management on either one or the other of change and complexity. Management could choose, metaphorically, to throw a switch and emphasize change, agility, and business alignment, or take the conservative line suited to more complex architectures and choose a plan-driven approach.

Now that the enterprise is essentially a dynamic and complex system, it is faced with managing both change and complexity. And as these twin challenges manifest and become inextricably coupled, the management gap identified in the fourth quadrant is exposed, indicating the need for new, innovation management approaches that can transcend both agility and planning. Such innovation is not of the traditional type that gives rise to new products and services, but process innovation that can improve our ability to cope with the challenges like those found in the enterprise pattern.

9.1.2.1 Projects as a Diminished Paradigm

Even as business and technology architectures have changed, our management approaches remain stuck in time. Practitioners continue to be training to behave as if each new project is a blank page of endless design possibilities. This textbook scenario is increasingly unrealistic and in many organizations, projects instead have to be managed within the constraints of an existing architecture. Attempts to scale the project paradigm to programs and portfolios focus only on the activity dimension without addressing the design complexity associated with the execution of multiple, interdependent initiatives. As such, these attempts stretch the existing paradigm but offer few new insights, with many of the guidelines, for example on managing dependencies, remaining vague and heavily reliant upon a manager's communication skills, experience, and other tacit knowledge.

Ongoing attempts to adjust the traditional project management paradigm, including the redefinition of the project itself, show a gradual metamorphosis of best practices as they attempt to accommodate the reality of change. The Agile Manifesto took a more revolutionary approach to the status quo, challenging the ability to establish comprehensive requirements specifications and so follow a detailed project plan. However, a more fundamental assumption at the heart of the project paradigm went unchallenged; everyone continued to enshrine the customer at the apex of power, influence, and control in the project. That ensured that customers or their proxies, including sponsors and program or project managers, continued to focus their attention on individual programs with scant attention to the interdependencies between them.

The singular and all-powerful customer is at the heart of the project paradigm and, from systems engineering, to quality management, to agile methods and even contracts law; we have always assumed that we must deliver the customer's requirements to their satisfaction. In the enterprise and in the modern management vernacular, focus on the business customer translates into business alignment, making sure that all shared services groups such as IT help support and enable the business strategy. Few have been willing to challenge this fundamental assumption and question if it is appropriate, or even possible, in the federated, multidimensional system of systems that describes the enterprise.

Is it possible that we have been tackling the wrong problem? How do we defer to the customer's requirements when there is more than one, and they each have different requirements? Can it be that the management problem, like a virus, has mutated and transformed so it is now essentially different from what it was before? This could help explain why earlier remedies are less effective today and point to the need to adapt to address the new management reality.

9.1.2.2 Trending to the Fourth Quadrant

Even accepting that traditional project management is less suited to the enterprise pattern, the question arises of how commonplace that situation is in reality. The Bain study provides an indication that only 10% of enterprises have successfully juggled both efficiency and business alignment. But we can expect that there will be a greater need, and considering the business benefits the study identified, a drive to master the challenges found in the fourth quadrant that is home to the enterprise pattern. If nothing else, the simple logic of the change-complexity cycle described in Chapter 2 will propel an increasing number of organizations into the top-right quadrant. Today's management challenges leave them little choice but to embrace both change and complexity.

Figure 9.1 shows the different paths organizations can take that lead inexorably to the top-right quadrant where the project paradigm is least successful. In Path A we can see a small agile team focused on business alignment

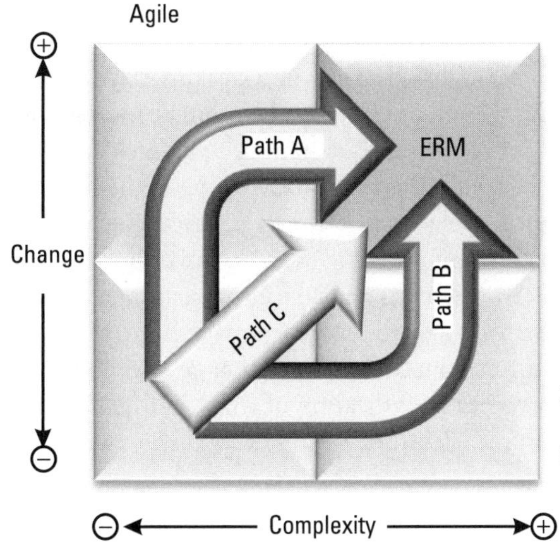

Figure 9.1 Pathways to the enterprise pattern.

developing solutions that meet the immediate requirements. The short-term expediency that drives this path ignores the investment needed to operationalize new solutions. That investment includes integration with business processes, staff training, and interfacing with other systems can make even minor solution components an integral part of the enterprise architecture that cannot easily be withdrawn from service. A point solution can be successful if it continues to work in isolation (as a silo) in that environment, but as soon as there is a need to integrate the stand-alone solution with other business components, it literally turns the corner to enter the fourth quadrant. When it is made to utilize existing infrastructure or middleware technologies,or in a non-IT example, the new capabilities are incorporated into existing business processes, complexity increases and management has to address interfaces and dependencies on external systems. This is one traversal of the change–complexity cycle, where simplicity soon gives way to complexity as point solutions need to be integrated into the operational environment. But equally the increasing need for the business to adapt to competitive pressures or quickly grasp new opportunities can drive enterprises with already complex and interdependent architectures to embrace change. Path B in Figure 9.1 represents this situation that is enthusiastically adopted by many larger enterprises. But the change and complexity cycle cannot be easily avoided and tends lead to the inevitable situation where teams can no longer act unilaterally and have to recognize they are part of the larger enterprise architecture. Finally, there is Path C, where rapid growth or expansion can catapult an organization from simple textbook scenarios of solution development into relative confusion of the enterprise pattern.

Starting with a focus on change and business alignment, an enterprise generally aims to increase or maintain revenues while managing complexity. Standardization, consolidation, and internal efficiency is geared to reducing costs this has the effect of limiting solution options. These forces represent two competing strategies which, put bluntly, are to make money and to save money. Obviously, mastery of the fourth quadrant can pay dividends when both can be accomplished, as shown by the Bain study in the IT management. These twin business drivers can be related to Porter's generic strategies of cost leadership and differentiation and impose a logical order—you have to first make money in order to save it. So the paths to the fourth quadrant have a relationship to maturity of an enterprise and the market in which it operates. This goes some way to providing an explanation of how change and complexity can arise in the business and provides an indication of when in the business cycle the ERM becomes most relevant. Clearly, elephants seeking to dance represent one class of enterprise that should adopt ERM practices—but another might be nimble dancers seeking to become elephants.

9.1.2.3 Management and the Release Paradigm

This book makes the argument for a new management framework centered on the release paradigm, as applied at the enterprise level and taking an end-to-end life cycle perspective. Presenting a holistic view of enterprise transformation and change, this release-centric approach can offer C-level executives and senior management visibility of the enterprise enabling them to better govern strategic execution and delivery.

The release paradigm provides a consistent matrix notation that represents the multiple customer–supplier relationships as they traverse the life cycle from inception and planning through to execution and delivery. Suitably applied, this notation consolidates critical management information at an appropriate level of abstraction that does not bury critical dependencies in the detail. As a result, it offers a reporting and tracking tool across the life cycle that can inform decision makers and provide greater insight into all changes that impact the enterprise.

The release paradigm offers a unifying framework for the different disciplines as they apply today, stitching them together and focusing them on the delivery of the strategic outcome. Figure 9.2 provides a schematic representation of how current management practices can be siloed, applicable only in specific phases of the life cycle. This management mosaic, while worthy of further discussion, is only illustrative and aims to show the belated focus on release management—often without clear linkage to the corresponding release planning mechanism at the start of the life cycle. The patchy application of the different disciplines across the life cycle can result in discontinuity and

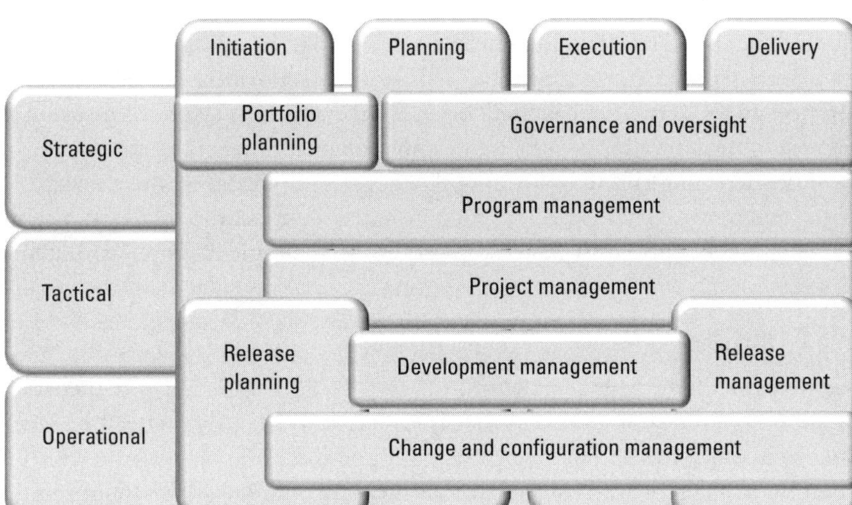

Figure 9.2 Management mosaic across the life cycle.

friction. Progress is reliant on effective interactions and handoffs between these disciplines yet different methods and notations can complicate communications and slow progress. Having instead a consistent notation and conceptual model across the end-to-end life cycle offers a shared perspective and a common working language, or lingua franca, for the delivery of a strategic change in the enterprise.

However, ERM goes beyond providing a new life cycle model. In recognizing and managing continual change, it aims to establish a framework that expedites the execution and delivery of changes by having preplanned and architected changes ready and waiting for the point where implementation resources become available. While this is a classic application of the front-end loading principle, the investment and preparation required is not only early in the release cycle, but actually outside of these individual cycles, instead providing a framework for planning and delivering the enterprise roadmap.

ERM offers this release management framework, designed to execute and deliver a change portfolio via reliable, timeboxed releases that are analogous to continuous build and integration in software development. Indeed, it is the experiences in software, from the agility of iterative cycles and the ability to manage multiple, concurrent changes that have provided the blueprint for the release paradigm scaled to the enterprise. Similarly, a range of supporting tools are envisaged to ease the application of ERM and allow the different parties and stakeholders involved in all stages of the release process to give and receive status, sharing pertinent information. Reviews of existing tools and prototypes of collaboration software suited to ERM can be found at EnterpriseRelease.com.

Clearly, the expansive scope of the release paradigm and the multiplicity it aims to address results in a more general and necessarily complex life cycle model. The natural aversion to complexity is mitigated by the relatively intuitive, visual representation that the matrix notation provides. Further, with the appropriate tools, different filtered views of the matrix life cycle can display the relevant information for specific parties, shielding them from the complexity of the other dependencies yet informing them of the implications where necessary. At the end of the day, it is reality that is complex, not just the model. Einstein's saying is apt in this context: Make everything as simple as possible, but not simpler.

9.1.2.4 Establishing Theoretical Foundations

Today, particularly in management, no one likes to use the word theory. It conjures up an idea or approach that is impractical (as in theoretical) or idealistic (as in ivory tower). Yet our perception and understanding of the world is based on theories or mental models that provide a framework to interpret our world. We make reasonable assumptions about cause and effect, stimulus, and response that are for the most part borne out by observation and experience. If you were to drop this book, it would be reasonable to expect it to fall down—we don't have to know of Newton or his theory of gravitation to understand that.

Once internalized, these assumptions become presumptions that we rely on and forget to test, even when conditions change. These assumptions become so much a part of our thinking that we become reluctant to give them up, and even make excuses for them when they no longer work[1]. Challenging an entrenched paradigm can appear to be heresy, especially when current theories continue to hold in the majority of circumstances. In management, as in science, it is at the boundaries of our knowledge that we begin to see signs of a failing paradigm.

The contention at the heart of this book is that projects are no longer effective as a delivery vehicle in the enterprise. The conditions under which this enduring paradigm fails can be found in the enterprise pattern. While the Bain study suggests that this may apply to a relatively small proportion of organizations, the pattern was not given its name lightly. Problems with the project paradigm (not simply project failures, but problems of representation and reporting) can be expected in modern enterprises of even moderate scale. Earlier arguments make clear that this can be expected to be a growing trend, given it taps into some fundamental forces at work in business and society at large.

As a result, whether we recognize it or not, we are in a time of shifting paradigms. Now is the appropriate time to review existing management theories to ensure that they model the current reality. Perhaps the reason theories, especially in management, have a stigma is because we so often have bad theories, ones that are idealistic and tell us how things should work instead of how they do work.

In introducing ERM, an attempt has been make to provide a logical argument for its relevance. As far as it is possible in the subjective world of solution delivery models and management practices, the arguments in this book represent a derivation of sorts. In an effort to ground ERM and provide it with a theoretical foundation, its key assumptions been the subject of research and trials, albeit in the more constrained domain of product line architectures. This research supports over two decades of the author's consulting experience, working with executives and managers seeking to implement management tools, and best practices in businesses around the globe. Much work remains to achieve academic rigor, but the case for ERM already has more credibility than most management approaches. The reasoned argument made in this book that led to ERM can be summarized as follows:

1. Albert Einstein, who supplanted Newton's theory of gravity with general relativity, notoriously questioned the probabilistic nature of the newer quantum mechanics with the statement: "God does not play dice."

1. Change and complexity are identified as dual forces in the modern enterprise;
2. The different management approaches that deal with each were discussed and a gap identified where both were present;
3. The enterprise pattern was proposed as a means of representing multiple changes that impacted interrelated components;
4. This pattern was represented as a simple data model that could be normalized to form a tabular or matrix representation for each release;
5. Several matrices could then capture pertinent data across the release life cycle providing a means of a consistent means of describing multiple concurrent delivery streams;
6. Coordination of activities across the different suppliers required synchronized baselines that created shared inflection points in the life cycle.

As a new model or theory for managing strategic execution and solution delivery, it must be stressed that ERM does not negate preceding management approaches. Instead, it offers a unifying model and generalized notation that can incorporate earlier life cycle models. The good thing about considering management theories is that we do not have to prove the past wrong, but simply identify the specific conditions under which they apply. While there are distinctive features in ERM that differentiate it from the project paradigm, practitioners may find it different but not alien. For those who can identify the enterprise pattern in their own organization, the question that remains is: How does an organization adopt or transition to benefit from ERM practices?

9.2 Adopting ERM Practices

Adopting ERM practices is a journey and it should, like any change that impacts the enterprise have a viable business case. That business case should be based on a realistic plan that establishes the adoption timetable, focusing on the quick wins that the matrix notation can provide practitioners.

Each specific plan for implementing ERM will depend on the culture of the organization and its readiness to adopt a release-centric paradigm. An organization used to strong centralized control will look to adapt the ERM differently to one where a more agile and innovative culture prevails. As such the following is only an outline indicating possible activities to be conducted in a sensible sequence. Those on the ground should consider their options and determine what is likely to work for them.

There is only one cautionary note to be sounded about the implementation of ERM that is obvious but needs saying. It is simply that while any change is difficult, it is particularly difficult to depart from the project paradigm that has been core to managing transformation and change in the enterprise. This book has taken pains to step through what we believe is irrefutable logic and evidence that points to the need for ERM. Clearly, all those in the organization who are asked to go on this journey and make the necessary adjustments to their own "mental models" should be provided sufficient exposure and training to ERM to come to an understanding of what it means to them and the organization.

9.2.1 Implementing ERM

The successful implementation of any new management process depends on there being a clear motivation for that change. To embrace ERM executives and managers need to see the potential benefit and be prepared to champion its adoption. These benefits, like those of best practices fall into two broad categories: improving effectiveness and efficiency in the execution of change programs, versus the reduction of delivery risk and the cost of quality. Simplistically, the options are to plan better or react faster to problems: be prepared or be agile.

The two key points at which ERM can be introduced in an enterprise reflect these options. We have the ability to pursue a top-down implementation of ERM or the more familiar bottom-up approach. In both, the matrix notation is valuable and offers a means to eventually establish an end-to-end release life cycle. But aside from specific implementation steps, ERM constitutes a process change, a departure from the status quo and the presiding, highly entrenched project paradigm. As such, implementing ERM requires a level of organizational change to at least explore the possibilities it offers. The generic steps the implementation of ERM involves are presented, but while the benefits of ERM are clear, this is the start of a journey. While the concepts in this book aim to be practical they are not complete—the release paradigm is still new and there remains much to be explored. Those interested in ERM will discover like-minded individuals, sharing experiences and ideas on EnterpriseRelease.com where additional implementation notes can be found. Outlined here are just the first steps towards ERM, but they are important in reframing enterprise change and transformation so it is more amenable to a solution.

9.2.1.1 Top-Down and Bottom-Up ERM

As a representation of multistreamed life cycle, ERM has been shown to have the greatest influence at two inflection points—the transition from release planning to execution and then from execution to delivery. At the first point, ERM calls for the consolidation of multiple solution architectures into a compatible, and agreed release architecture. It is this architecture that guides the separate component teams' implementation efforts, recognizing that the partial deliveries of multiple solutions from the projects still have to make a coherent whole—the release. At the second inflection point these components then have to finally deliver their capabilities that must be integrated together and tested before the release is finally implemented.

The implementation of ERM in an organization therefore must focus on supporting these twin milestones where the multiple project life cycles have to be synchronized to coordinate the efforts of the multiple parties—these are the two critical baselines for the release. As the first of these relates to a top-down implementation of ERM, where the business planning and architectural design takes a release-centric view before the different component groups can fan out to implement the changes needed. Applying ERM principles to the second inflection point takes the alternative bottom-up approach that focuses on release integration and test, taking a more quality control and risk-reduction perspective. Clear governance gates must be established at each of these two pivotal points in the life cycle if an end-to-end implementation of ERM, one that delivers the greatest value is to be achieved. However, at these two ends of the life cycle there the stakeholders that need to subscribe to a release-centric approach are quite distinct.

Top-down release governance has to become a part of portfolio prioritization and introduces new steps in the inception and approval of new business initiatives. Tackling the issue of uncoordinated changes at its source can appear relatively simple as in most organizations the tracking, if not the allocation, of project funding is already centralized in one or more PMOs. The deceptively simple change that ERM requires at this level is to coordinate the funding so that initiatives are approved as a tranche with execution planned for a specific release timebox. However, that presupposes a release-centric framework is in place so top-down implementation of ERM must necessarily focus on the following:

- Establishing an enterprise roadmap that identifies how the strategic vision is to be delivered in forthcoming release milestones or timeboxes. This roadmap sets the expectation for upcoming enterprise change, even where they are not yet fully defined, over a timescale that needs to reflect the volatility of the business environment. This can either simply periodic or based on predictable business milestones but needs to be a realistic and achievable. Where a regular cycle of releases is defined it needs to be a "Goldilocks" frequency that is neither too rapid nor too slow for the different lines of business as well as the core component teams responsible for delivery. While changes can still be expected outside the roadmap, especially at the early stages of adoption, the roadmap must at least cover the core business systems that are shared across the enterprise.

- Establishing a new level of release governance that can be personified in the assignment of an architect for each of the releases identified in the enterprise roadmap. This release architect role acts as a design authority responsible for the release's scope and architecture, and is supported by business and technical specialists as required. The theme of each release in strategic roadmap needs to be fleshed out and the release matrix can capture the selected changes that define the scope of a coherent release with clear impacts and dependencies on components of the enterprise architecture. The change portfolio to be expected in each release must take as its planning baseline not the current state or as-is architecture, but rather the to-be state of the pervious enterprise release. In this way each release is planned to build on the previous one providing the basis for a reliable enterprise evolution.
- The strategic portfolio prioritization processes that operate in the enterprise do not have to, but may change as a result of implementing the ERM constructs above. While the release-centric focus is principally about the execution of strategy and not the quality of that strategy, there can be expected to be adjustments necessary in the portfolio prioritization selection process. The role of release governance is to identify incompatible or conflicting change initiatives and that can be expected to push back on previously unchallenged strategic programs. The need to rescope or replan these initiatives inevitability results in dialog that dependent upon the power of the release governance step. In fully embracing a release-centric perspective change initiative can be expected to be simply an input into an enterprise architecture that is then incrementally delivered as releases. This is the purist view of enterprise transformation that assumes centralist design control can be asserted. What is more likely is that the powerful stakeholders who initiate change programs will be advised by the release architect of the risks and additional costs associated with the execution of their current vision, with options and recommendations to have it better fit with the enterprise roadmap and more specifically the architecture of upcoming releases that the initiative targets.

The bottom-up implementation of ERM is the more familiar and essential aspect of delivery governance that is needed to ensure integration and testing of a change portfolio. The impacts of uncoordinated changes and defects quickly become evident in the enterprise and make it easier to convince management of the need to coordinate changes before they impact the business and its enabling systems and processes. As such, aspects of ERM are evident in most organizations and it is here that traditional change control and configuration management practices provide essential protection to operational system. And as the pace of change quickens in the enterprise, it is not long before the different IT and business systems demand a more regimented mechanism for introducing changes that see the beginnings of ERM at the second inflection point that aims to consolidate multiple changes so that they can be integrated and tested as a single release.

As discussed earlier, the bottom-up release governance that comes to focus at the delivery end of the life cycle is independent of the top-down release governance discussed above. In fact, the bottom-up governance becomes all the more essential when the initiation of multiple projects is not coordinated or controlled. In such a situation, there is a need to contain and corral the multiple streams and force them to fan in so that these separate changes can be integrated and tested. ERM at the second inflection point in the life cycle can therefore be considered like a lens that collects and focuses the separate changes to a single release. Such supply-side governance has to be put in place if only to protect the reputation of service functions like IT. While essential, without the visibility or influence over the early inception stages ERM becomes more like quality control rather than quality assurance—there is a need now to test in quality rather than build it in from the outset. There can be less resistance to this latter point of release governance because no project, no matter how independent its sees itself, wants to fail at the point of delivery even if that can be attributed to the interference of other projects. As a result the bottom-up implementation of ERM takes the more acceptable form requiring the following measures:

- A release calendar needs to be established, similar to the enterprise roadmap of the top-down approach but with the critical difference that the release milestones and timeboxes that are established are a reactive measure rather than a strategically planned roadmap. Nonetheless, there is a need to engage business sponsors to establish the frequency of enterprise releases as it relates to the required capabilities being provided by core component teams—even though ERM at this point can take no responsibility for what is actually delivered. While business customers and component suppliers agree the release calendar, bottom-up ERM governance is only able to administer delivery schedules and coordinate the integration and testing activities. The capability offered by the release itself is a black box (except for the defects that might be raised) and projects remain responsible for the quality of the contents.
- The participants of each release have to be determined and identified prior to commencing integration and test phase. Identifying the partici-

pants entails establishing a release matrix with those projects expecting to have capabilities delivered and the component teams that have accepted the allocated requirements. This effectively determines the scope of the release from a delivery perspective rather than what might have been planned. In the absence of top-down ERM, the project teams are responsible for any discrepancy between the planned scope and delivered scope. All that bottom-up ERM can address is integration and testing of what is to be delivered in that release.

- Handover and migration processes have to be clear from the component teams to the integration and test teams. For IT systems this entails the establishment and management of test environments that have to be configured with the latest component releases. This is a classic configuration management capability that also needs to have adequate change controls as raised defects are fixed, resulting in new versions of the components being brought into the test environment. The non-IT equivalent is equivalent where business processes must be walked through and validated before user acceptance marks the final implementation of the release and the migration of IT components into the production environment.
- Reporting the release status as it undergoes integration and testing is the most visible and crucial stage of ERM as the business prepares for the necessary transition. Clearly, this is a point of reckoning where previous agreements and expectations have to be continually ratified, making the governance around the release matrix a key exercise that involves all parties through involved in the final delivery stages. Here all projects need to reconcile what was promised with what was finally delivered—and successfully passed the integration and testing. At the same time, the component teams have to address raised defects, resolving those they can and negotiating the severity of those that require more time than the release schedule permits. Ultimately, the decision to accept the release into production has to be made on the basic of rapidly changing information when agreements between the different teams have to be reassessed and updated. It is this point that the matrix notation can be applied to great effect as it offers the framework within which the parties directly involved as well as any consequential impacts can be determined.

Clearly, it is the goal of ERM to achieve the strategic delivery of a change portfolio to become a part of portfolio prioritization and introduce new steps in the inception and approval of new business initiatives. Tackling the issue of uncoordinated changes at its source can provide greater certainty for the larger, strategic initiatives (the big rocks) however the volatility in most businesses can be expected to require a robust bottom-up release governance to match the top-down strategic planning application of ERM.

At which every point that ERM implementation is commenced the relevance of the release life cycle should soon become clear and the application of a release-centric perspective should gradually spread, upstream or downstream, to eventually provide end-to-end coverage. And the unifying matrix notation has relevance equally at all stages, able to provide a comprehensive definition of the release at each stage, with the very dimensions of the release matrix providing an outline of the release architecture itself.

9.2.1.2 Generic ERM Implementation Plan

The successful implementation of any new management process depends on there being a few motivated executives or managers who see its potential benefit and are prepared to champion its adoption. In terms of ERM, the transition from the powerful project paradigm to a release-centric one is best undertaken incrementally, letting ERM supplement the information that projects already provide. There does not need to be an overthrow of the existing order and a revolt against project-centric things because its limitations are usually already understood by practicing managers. Instead a consolidation of existing project information into the release perspective can naturally lead to the realization of ERM's value—where it is found relevant.

Assessment

The simplest way of assessing the relevance of ERM and generally mobilizing the organization is to undertake a review and assess its strategic execution capabilities. Along with other benchmarks, the relevance of enterprise pattern in the organization can determine its readiness for ERM processes. Questionnaires that aim to assess the engagement of the different disciplines in the release cycle also provide a gauge of the value of ERM. Samples are available on EnterpriseRelease.com for download and offer a quick assessment of the contribution that a release-centric perspective can provide the different roles across the delivery life cycle.

Of course if an assessment shows that the organization does not have the challenges found in the enterprise pattern, and is working successfully using either the textbook or integration patterns, then needless to say, ERM should only be adopted if a change in conditions is anticipated. In any event, a new paradigm cannot be forced, it can only be recognized. And once recognized to be relevant, there is inevitability to its adoption even as the familiar change management approach can speed transition to the new paradigm.

Where the organizational assessment determines if an enterprise has the challenges ERM attempts to meet, it is also a barometer for change providing a measure of the organization's readiness to adopt ERM. It should also provide a perspective on what the key stakeholders participating in the assessment consider the strengths and weaknesses of the organization. Clearly, the project management community, along with the enterprise architects and the component development groups, will be the most interested (and sensitive) to the introduction of ERM.

Assuming then that such an assessment is conducted and shows that the ERM has relevance, the following steps provide the outline a generic implementation plan that can be reviewed and appropriately tailored for each organization's unique conditions.

Mobilize for Change

The ERM implementation approach needs be sensitive to the strengths and weaknesses discovered in the initial assessment. Mobilizing the organization for change can benefit from focusing ERM on the weaker disciplines and processes identified. A series of workshops with key representatives from the business, projects, and functional groups like IT can participate in a two-way discussion of ERM principles and constructs. This provides the opportunity to introduce ERM and explore the implications on its implementation on existing roles.

ERM needs to be understood as a unifying perspective that emphasises the release-centric perspective without requiring existing processes be fundamentally changed. Issues to address as part of the mobilization for ERM should confront potential points of resistance. Where an organization has a strong project management capability, some opposition to the idea that a project or program might be subordinate to a release can be anticipated. The collaborative nature of ERM can be stressed as a valuable supplement for project controls already in place. Where a strong collaborative culture might exists, the matrix notation can be used as an exercise in establishing a shared a consolidating perspective that highlights external dependencies.

The fundamentals of ERM such as the twin inflection points present topic for discussion and related to existing techniques currently applied by the team at those points. Similarly, the introduction of release governance is an area with multiple implementation options. Assignment of an individual in the release architect role may not be appropriate in some organizations preferring shared responsibilities. Governance may be more appropriately implemented with a cross-functional team. And if so, which roles should be involved? While senior program managers and business leaders can be seen as candidates in such a governance body, every enterprise can be expected to have its own views on their implementation—the ERM simply insists that such a role or body exists.

Making the ERM concepts relevant to a specific organizational context while identifying any differences to current practices allows a discussion of the value of particular aspects of ERM and gives each construct an implementation a priority. The manner in which these are implemented can be the subject of subsequent ERM design workshops held with those sponsoring the initiative.

Plan a Future Enterprise Release

The best way to understand that changes that will be needed to implement ERM is to start planning a future enterprise release. By selecting a date that is past the current work in progress, the organization has the ability to reset the release cycle and define the release roadmap beyond that point.

The activities necessary to mobilize for ERM and plan the first enterprise release can themselves be managed as a pilot project, aimed at introducing and adopting ERM in the organization. The focus on an actual future release date can expedite the application of ERM and prevent it from being an academic exercise. Only a small core team is necessary to coordinate the release planning and establish templates for the release matrices that define and document its life cycle stages. The core skills of these individuals relate directly to the dimensions of the matrix itself and should include one of each of the following:

- Executive sponsor seeking to improve visibility and insight of enterprise change and transformation;
- ERM evangelist to facilitate and guide the planning process;
- Portfolio or program manager familiar with the current and upcoming releases;
- Enterprise architect with knowledge of the business and technology components impacted by the change;
- Change manager to identify the organizational and cultural changes;
- Business/process analyst(s) to document the requirements of the future release and the ERM procedures to be used for future releases.

The team should essentially undertake the design of ERM processes and trail the matrix notation. This can first entail the definition of the projects participating in each release and the component teams delivering to their needs. This essentially frames the specific release and determines the "release team" focused on a specific deliverable in the enterprise roadmap. As a pilot of ERM, the wider concerns relating the creation of such a roadmap can await the full adoption of ERM, for which the pilot can be expected to provide useful guidance.

At the inception stages of the release, the two concepts that should be trialed are the establishment of a unified change portfolio and the establishment of the release architecture—both described below. Experience suggests that a key part of the pilot will be establishing these constructs at a suitable level of detail and abstraction. The tendency that will have to be avoided, and actively resisted at times, is to go into too much detail in describing these. The value of a breadth-first approach is not usually seen; however, the restrictions of the matrix notation itself should enforce some brevity. The aim that must be pursued is the description of capabilities in business language and a solution architecture with enough detail to allow estimation by component teams. This is of course a slippery slope and while assumptions need to be made, they must not cloud the simple language needed to communicate meaningfully with business stakeholders.

Extending the use of the matrix notation across the life cycle not only captures the data critical for each phase, it also guides the processes needed to collect it and identifies the roles within the organization that have to be engaged in the release process. Particularly at the inflection points at the release design and integration the interactions that arise around the matrix can offer a frame of reference for both what is done, as well as what should be done to coordinate the multiple streams and reduce the risk undetected dependencies.

Planning for a distant future date, the negotiations between parties may not have a sense of urgency; however, if the change portfolio is maintained and priorities are established it can be surprising how stable a plan can be developed. The reality is that this plan will be adjusted as the date for the release approaches; however, with the suitable identification of risk the goal is to establish realistic expectations that the estimates and designs can and will evolve with the solution are elaborated and be the subject of future (re)prioritization and (re)negotiation as required. As in agility, the plan itself needs not be the end in itself, but rather a guide that used together with the new collaborative processes that ERM enables (and should be identified in the pilot) can ensure a more optimum release.

Established a Unified Change Portfolio

Creating a plan for a single future release is in fact a misnomer and while planning a single enterprise release, longer-term implications inevitably arise. If a project cannot be fully implemented in the current release, where are the outstanding capabilities that cannot be addressed recorded? And, presuming they remain relevant priority, when can they be expected to be delivered? These are of course longer-term concerns that can be expected to be raised by the pilot even it has a focus on a single future release. However, if ERM is to be adopted these fundamental questions must be answered and the necessary tools and infrastructure established.

Establishing the scope of any enterprise release cannot be expected to be simply determined from new initiatives not previously considered. Projects that are already underway can be expected to have future releases planned and these have to be considered in planning any enterprise releases. With the prevailing project perspective a consolidated view of the backlog of initiatives and their scope is not always easy. While the PMO(s) may have the project names and budgets they are less likely to maintain a clear definition of their scope. Further, the PMO is primarily concerned with the scope that was approved, not those initiatives that were rejected or deferred. So it is only in more mature organizations that a list of prospective initiatives and their scope item can be expected to be maintained, even though there is nothing particularly difficult about doing so.

The establishment of such a repository of initiatives and scope items from across the enterprise is not a technical problem—it is an organizational one. It is necessary for ERM because simply accepting and planning initiatives as they are proposed in a business case may not be effective from a release-centric viewpoint. There is a lot greater flexibility in planning a release when each project or its business case identifies capabilities and describes them in terms of more granular scope items. This is akin to feature-driven development and with scope defined a more granular level; release planning can optimize its delivery capacity. This is accomplished by mixing appropriate big rocks with smaller rocks to get the greatest value from each release. And to establish a truly unified change portfolio with fine-grained changes requests provides the most flexibility even if it is likely to require the consolidation of several change repositories from across the organization.

Establishing the existence of a consolidated and unified change portfolio that records the backlog of scope items, separate project plans, and even change requests shows the change forces acting on the enterprise. While these changes need to be suitably categorized and prioritized, with the relationships between them maintained, the granularity offers greater flexibility the release architect in planning for their execution. This is not to imply that all changes receive the same level of review and design attention. Priority scope items with strategic import clearly must be allocated to releases ahead of lesser priority changes and indeed should set the theme for entire releases. The value of smaller changes being considered is that they can more easily be accommodated in a coherent release architecture but that should not overturn the need for business priorities to be met.

A means of establishing a view across all changes in the enterprise is the ultimate goal, whether implemented as a single repository or a consolidation across multiple tools. The decomposition of projects into smaller scope items allows a democratic change process that can support innovation and allow greater autonomy for those implementing the changes. As previously discussed, aside from the identified priority requirements, it is envisaged that teams can choose to implement selected smaller granularity changes either as stretch targets or to fill any delays resulting from interdependencies. The feasibility of such innovations needs to be assessed although there is little to lose as the implementation amounts to little more than making the backlog visible and ensure that the constraints of planning do not artificially constrain the work that can be performed.

An Actionable Enterprise Architecture

Experiences from the pilot can be expected to adjust the focus of some disciplines and in particular make the work of solution and enterprise architects more focused and actionable. The release focus ensures that architectural baselines are seen in the evolutionary context with each release laying down the foundations for the next release. The emphasis on these transition architectures over the more idealistic future-state blueprints or the equally artificial solution architectures that only represent the final state consumes significant effort and distracts from the reality of the intermediate architectures that are actually delivered.

A release architecture is essentially an actionable enterprise architecture targeted at a specific point in the enterprise roadmap. Simply establishing the columns in the release matrix requires the identification of the components impacted by the change initiatives, and goes some way to outlining that architecture, at a level which few architects would know of today because they are not currently engaged at such a practical level. Therefore, even the apparently simple act of identifying the component teams involved in a release can be expected to generate debate and discussion. Mature architectural practices can be expected to have an agreed taxonomy for the architectural components themselves, but in many organizations where architectural standards have not been established, determining the release matrix beyond the organizational units involved can be a catalyst for exposing the need for shared naming conventions.

The different levels of architectural decisions can benefit from the exposure that the matrix notation provides. Enterprise architecture in particular can be seen to be directly relevant to planning for change instead of being viewed as an esoteric discipline. Instead of being buried in design documentation, the critical decisions made by architects (such as determining the components to be modified to implement a new business capability) are captured in the rows of the release matrix and made clear to all parties from the outset of an initiative. Further, the need to consolidate the separate solution architectures and ensure that the individual component designs can accommodate multiple change initiatives that impact them becomes intuitively obvious in the matrix notation. Finally, the release matrix as a whole makes clear that all the individual design decisions together represent transition architectures for the enterprise as a whole. When agreed and baselined, this release architecture must form a stable future state of the enterprise and any incompatible or inappropriate design decisions can directly impact upon the business. This makes clear the need for the role of release architecture in the initiation and planning stages of the release, where the review and consolidation of the separate solutions designs can ensure the delivered as an integrated and low-risk release.

9.2.1.3 The First Exercise

The relevance of the concepts in this book is best explored in discussion with colleagues and with real-world scenarios. The intuitive nature of the release matrix can be used as a starting point to understand the different ramifications that can take pages to explain—they are better understood through play and experimentation.

If you have already done so, attempting to create an outline release matrix for your organization or project context can be most instructive. This can be on done on a whiteboard with trusted colleagues or even on the back of an envelope or a napkin over lunch. A sketch of the release matrix with the relevant projects and their impacts on the component teams involved can help to gauge the relevance of the concepts without recourse to overloaded terminology that can obscure the pattern. That then is the first exercise: to attempt to achieve a consensus of what constitutes the current release matrix in your organization. And from the discussion and feedback we are confident that a shared perspective of the forces shaping the future direction of the business can be determined and clearly communicated.

The matrix notation recursively asks the simple management question of each strategic initiative: To achieve this, who needs to do what? The answer to that simple question can populate the release matrix and define the transformations necessary to take the enterprise from its current state to the desired target state. As a recurring or self-similar pattern, it can also be instructive to consider the entities at both the higher-order strategic initiatives that give context to the current changes. Similarly, attempting to decompose the rows and columns of the matrix you have created can identify detailed level of the release, identifying the groups and individuals upon which the release relies.

At the end of the day, the matrix representation can only reflect your change portfolio and the architectural components that are impacted; it cannot make sense of them—that is your contribution. As a communication and collaboration tool, the release matrix does not require well-designed change programs or elegant, componentized architectures. It is simply a management tool that helps to make explicit dependencies that might otherwise be ignored.

Finally, I am encouraging you to share your experiences online at EnterpriseRelease.com. There you will find the latest resources on ERM including questionnaires and tools to help your team quickly develop a release-centric management framework. If this book has been successful in stimulated thinking on the adequacy of the existing project paradigm and sparked an interest in exploring the new release paradigm, then do please join the discussion. There is still a lot of work to do!

References

[1] Taborda, L. J., "Planning and Managing Product Line Evolution," *Proceedings of the 5th Product Family Engineering (PFE-5),* F. van der Linden, (ed.), New York: Springer-Verlag, 2003, pp. 296–309.

[2] Taborda, L. J., "Generalized Release Planning for Product Line Architectures," *Proceedings of the 3rd Software Product Line Conference (SPLC 2004),* R. L. Nord (ed.), New York: Springer-Verlag, 2004, pp. 238–254.

[3] Taborda, L. J., "The Release Matrix for Component-Based Software Architectures," *Proceedings of the 7th International Symposium on Component-Based Software Engineering (CBSE 2004). I. Crnkovic, et al.,* New York: Springer-Verlag, 2004, pp. 100–113.

[4] Gerstner, L. V., *Who Says Elephants Can't Dance? Inside IBM's Historic Turnaround,* New York: HarperCollins, 2002.

About the Author

Dr. Louis J. Taborda is an international consultant working at the boundary between business and IT management. He has wide industry experience implementing industry best practices and holding senior roles in software, engineering, and program management. Dr. Taborda also held the position of Director Lifecycle Management at a leading software tool vendor where he helped develop the vision for an integrated application life cycle. Today, he evangelizes an end-to-end approach to solution development and delivery that improves productivity and reduces risk. Working with clients across the United States, Europe, and Asia, he helps organizations streamline their release management and governance processes.

Dr. Taborda was awarded a doctorate for his research into the evolution of complex architectures and the identification of new release management tools and techniques. He remains active in the IT community and operates the online magazine Alinement.Net and EnterpriseRelease.com, a community site aimed at exploring the application of enterprise release management. He has a B.Sc. (Hons) from the University of Western Australia majoring in physics, with graduate diplomas in computer science from the University of Sydney and education from the Australian Catholic University. He was granted his Ph.D. from the Macquarie Graduate Management School in 2007.

Index

Recent Titles in the Artech House Technology Management and Professional Development Series

Bruce Elbert, Series Editor

Actionable Strategies Through Integrated Performance, Process, Project, and Risk Management, Stephen S. Bonham

Advanced Systems Thinking, Engineering, and Management, Derek K. Hitchins

Critical Chain Project Management, Lawrence P. Leach

Decision Making for Technology Executives: Using Multiple Perspectives to Improve Performance, Harold A. Linstone

Designing the Networked Enterprise, Igor Hawryszkiewycz

Engineering and Technology Management Tools and Applications, B. S. Dhillon

The Entrepreneurial Engineer: Starting Your Own High-Tech Company, R. Wayne Fields

Enterprise Release Management: Agile Delivery of a Strategic Change Portfolio, Louis Taborda

Evaluation of R&D Processes: Effectiveness Through Measurements, Lynn W. Ellis

From Engineer to Manager: Mastering the Transition, B. Michael Aucoin

Introduction to Information-Based High-Tech Services, Eric Viardot

Introduction to Innovation and Technology Transfer, Ian Cooke and Paul Mayes

ISO 9001:2000 Quality Management System Design, Jay Schlickman

Managing Complex Technical Projects: A Systems Engineering Approach, R. Ian Faulconbridge and Michael J. Ryan

Managing Engineers and Technical Employees: How to Attract, Motivate, and Retain Excellent People, Douglas M. Soat

Managing Successful High-Tech Product Introduction, Brian P. Senese

Managing Virtual Teams: Practical Techniques for High-Technology Project Managers, Martha Haywood

Mastering Technical Sales: The Sales Engineer's Handbook, John Care and Aron Bohlig

The New High-Tech Manager: Six Rules for Success in Changing Times, Kenneth Durham and Bruce Kennedy

The Parameter Space Investigation Method Toolkit, Roman Statnikov and Alexander Statnikov

Planning and Design for High-Tech Web-Based Training, David E. Stone and Constance L. Koskinen

A Practical Guide to Managing Information Security, Steve Purser

Preparing and Delivering Effective Technical Presentations, Second Edition, David Adamy

Reengineering Yourself and Your Company: From Engineer to Manager to Leader, Howard Eisner

The Requirements Engineering Handbook, Ralph R. Young

Running the Successful Hi-Tech Project Office, Eduardo Miranda

Successful Marketing Strategy for High-Tech Firms, Second Edition, Eric Viardot

Successful Proposal Strategies for Small Businesses: Using Knowledge Management to Win Government, Private Sector, and International Contracts, Fourth Edition, Robert S. Frey

Systems Approach to Engineering Design, Peter H. Sydenham

Systems Engineering Principles and Practice, H. Robert Westerman

Systems Reliability and Failure Prevention, Herbert Hecht

Team Development for High-Tech Project Managers, James Williams

For further information on these and other Artech House titles, including previously considered out-of-print books now available through our In-Print-Forever® (IPF®) program, contact:

Artech House
685 Canton Street
Norwood, MA 02062
Phone: 781-769-9750
Fax: 781-769-6334
e-mail: artech@artechhouse.com

Artech House
16 Sussex Street
London SW1V 4RW UK
Phone: +44 (0)20 7596-8750
Fax: +44 (0)20 7630-0166
e-mail: artech-uk@artechhouse.com

Find us on the World Wide Web at: www.artechhouse.com

Systems Engineering Principles and Practice, H. Robert Westerman

Systems Reliability and Failure Prevention, Herbert Hecht

Team Development for High-Tech Project Managers, James Williams

For further information on these and other Artech House titles, including previously considered out-of-print books now available through our In-Print-Forever® (IPF®) program, contact:

Artech House
685 Canton Street
Norwood, MA 02062
Phone: 781-769-9750
Fax: 781-769-6334
e-mail: artech@artechhouse.com

Artech House
16 Sussex Street
London SW1V 4RW UK
Phone: +44 (0)20 7596-8750
Fax: +44 (0)20 7630-0166
e-mail: artech-uk@artechhouse.com

Find us on the World Wide Web at: www.artechhouse.com